Introduction to
Nanotechnology

Introduction to
Nanotechnology

Massimo F Bertino
Virginia Commonwealth University, USA

World Scientific

NEW JERSEY · LONDON · SINGAPORE · BEIJING · SHANGHAI · HONG KONG · TAIPEI · CHENNAI · TOKYO

Published by

World Scientific Publishing Co. Pte. Ltd.

5 Toh Tuck Link, Singapore 596224

USA office: 27 Warren Street, Suite 401-402, Hackensack, NJ 07601

UK office: 57 Shelton Street, Covent Garden, London WC2H 9HE

Library of Congress Control Number: 2021949921

British Library Cataloguing-in-Publication Data
A catalogue record for this book is available from the British Library.

INTRODUCTION TO NANOTECHNOLOGY

ISBN 978-981-123-160-5 (hardcover)
ISBN 978-981-123-303-6 (paperback)
ISBN 978-981-123-161-2 (ebook for institutions)
ISBN 978-981-123-162-9 (ebook for individuals)

For any available supplementary material, please visit
https://www.worldscientific.com/worldscibooks/10.1142/12142#t=suppl

Desk Editor: Joseph Ang

Typeset by Stallion Press
Email: enquiries@stallionpress.com

To Ulrike, Anna and Laura, who keep the pot from cracking.

Preface

This textbook is conceived for a one-semester course at the upper undergraduate or freshman graduate level. The book was written with the following in mind:

- Nanotechnology is a vast field. It goes from paint to nanomedicine, through plasmonics and catalysis. An introductory course must compromise between quantitative and qualitative treatment. I tried to stay as quantitative as possible. I focused on the key physical and chemical principles and included theoretical treatment in an amount compatible with the one-semester time constraint. Depending on the make-up of the class, one may need to integrate the course with elements of solid-state physics or organic chemistry. I typically use excerpts of textbooks focusing on the subjects that I need to cover to bring the students up to speed.
- I tried to put things in perspective, stating the limits of the nano-approach, and tried to remove the hype that is frequently encountered in the field. A recent article, entitled *"Will any crap we put into graphene increase its electrocatalytic effect?"* (*ACS Nano*, **14**, 21–25 (2020)), should be required reading for students and practitioners in the field, including myself.
- In some cases, I shied away from work done by famous groups and included work done by not-so-famous ones. Sometimes, the work by the lesser-known groups is more clearly explained and better suited for a textbook. Also, in my view it is instructive for students to realize that there are a lot of competent people out there.

- There is no homework. When I teach this course, my assignments are tailored to the audience and are sometimes open-ended. I may give an article to read and ask the students to find the mistakes in the manuscript. Or, I may ask questions that I had to answer myself while preparing a lecture, or writing a paper. For example, one may assign homework asking why the thermal conductivity of He is higher than that of Kr, or why TEM specimens have to be so thin. The answers to these questions cannot be readily found on textbooks, or on the internet for that matter. Students who understand concepts and apply them will stand out. Students who are only good at memorizing will not do so well.

Massimo Bertino
Richmond, December 7, 2021

Contents

Chapter 1

"Trivial" Size Effects at the Nano-Scale

Nanotechnology is not necessarily new physics. In several cases, many effects at the nano-scale can be understood based on geometric considerations, or by simply applying basic physical concepts to determine what happens to a finely divided system. Some of these effects will now be described to show that, while they are not conceptually new, they do indeed contain powerful insights.

1.1. Surface area and surface-to-volume ratio

Let us consider a chunk of a material in the shape of a cube of side d. Assume to break the material in smaller cubes, each with side a. We want to know how many cubes there will be inside the cube, and how many of them will be on the surface. This is key to understanding the surface-to-volume ratio.

Looking at Figure 1.1, we can calculate that the total number N of small cubes is given by

$$N = \frac{d^3}{a^3}. \tag{1.1}$$

The number N_s of cubes with at least one face on the surface will be

$$N_s = \frac{6d^2}{a^2} \tag{1.2}$$

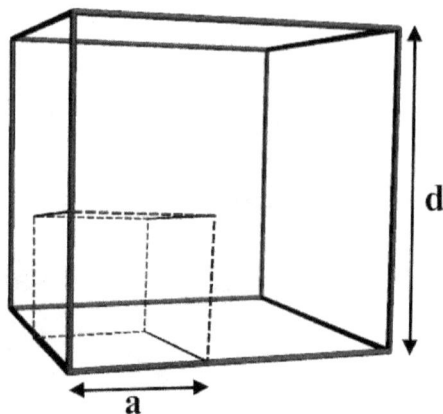

Figure 1.1: Splitting a cube of size d into cubic nanoparticles of size a.

and the surface-to-volume ratio S/V is

$$\frac{S}{V} = \frac{N_s}{N} = \frac{6a}{d} \tag{1.3}$$

(the cubes at the corners are a bit of a special case, but they are few anyway, so we ignore them).

Example. Assume to have $m = 1\,\text{g}$ of SiO_2 and assume to split it into cubes of $a = 20\,\text{nm}$ side. What will be the surface area after splitting? The density of silica is $\rho = 1.6\,\text{g/cm}^3$. The volume of the silica chunk is $V = m/\rho$. The volume of one small cube is $v = a^3$ and the surface of a cube is $6a^2$. The total number of cubes is

$$N = \frac{V}{v} = \frac{m}{\rho \cdot a^3}$$

and the total surface of the cubes is

$$S = N \cdot 6a^2 = \frac{6m}{\rho \cdot a}$$

Using the figures provided above, we calculate that $1\,\text{g}$ of SiO_2 yields 7.8×10^{16} nanoparticles, and a total surface area of about $180\,\text{m}^2$. So, splitting one gram of silica into $20\,\text{nm}$ cubes yields an area of about the size of a city apartment. We note here that high surface area materials are not just theoretical concepts. For example, Metal Organic Frameworks (MOFs) and aerogels have surface areas on the order of 500–1000 m^2/g. These high surface area materials have many applications, the most prominent being catalyst supports, filters and thermal insulators (see also Section 1.5).

1.2. Number of atoms in a nanoparticle

The following is also an elementary calculation, which, however, yields some surprises. Let us cosider a cubic (again!) nanoparticle of size d, so that we can use Eqs. (1.1) through (1.3). A typical nanoparticle has a size $d = 10$ nm, and in most metals the distance between nearest-neighbor atoms is the lattice constant a multiplied by $\sqrt{2}/2$ (this comes from geometric considerations of the face-centered cubic lattice structure). For a typical metal, $a = 0.4$ nm, hence:

$$N = \frac{d^3}{a^3} = \frac{10^3}{\left(\frac{0.4 \times \sqrt{2}}{2}\right)^3} \sim 4.4 \times 10^4 \text{ atoms}$$

$$N_s = \frac{d^2}{a^2} = \frac{10^2}{\left(\frac{0.4 \times \sqrt{2}}{2}\right)^2} = 7650 \text{ atoms}$$

$$\frac{S}{V} = \frac{N_s}{N} = \frac{7650}{4.4 \times 10^4} \sim 17\%$$

This example looks trivial, but it illustrates a common misconception. Specifically, we often encounter the notion that nanoparticles are special because they contain "just a few atoms", and that they have a very high surface-to-volume ratio. The example shows that a metallic particle with a size of 10 nm contains more than 40,000 atoms, and that less than 20% of these atoms are on the surface. We expect the thermodynamic and electronic properties of such a nanoparticle to be close to that of the bulk metal, which is often the case. Let us now consider a nanoparticle 5 times smaller, $d = 2$ nm. Then: $N = 350$ atoms, $N_s = 306$ atoms, and $S/V \sim 87\%$. For this particle, we do expect substantial deviations from the bulk properties!

To reiterate this point, the fraction of surface atoms is reported in Figure 1.2 as a function of the total number of atoms and of the size of a nanoparticle, assuming a lattice constant of 0.4 nm.

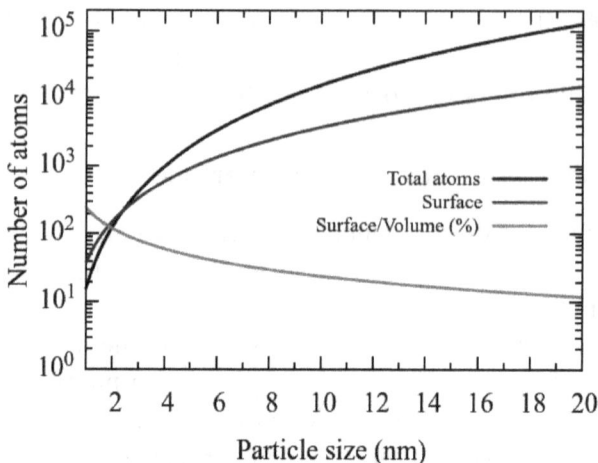

Figure 1.2:   Fraction of surface atoms as a function of the total number of atoms in a nanoparticle. An interatomic distance of 0.4 nm was assumed, and particles and unit cell were considered cubic. Note how Eq. (1.3) breaks down for very small particle sizes and yields >100% surface atoms.

1.3. Depression of the melting point

Equations (1.1) through (1.3) can also be used to learn quite a few things about the energy of a nanoparticle. The total energy E of a nanoparticle can be written as the sum of the binding energy of atoms (E_{bind}) minus the energy E_s necessary to create a surface in a solid. The energy to create a surface is proportional to the surface energy γ: $E_s \propto N_s \cdot \gamma \cdot a^2$, where a is the lattice constant and N_s is the number of surface atoms. The surface energy is the energy necessary to open a unit surface in a bulk sample and has a value $\gamma \sim 2 \mathrm{J/m^2}$ for most metals. For a particle with a total of N atoms and N_s surface atoms:

$$\frac{E}{E(bulk)} = \frac{N \cdot E_{bind} - N_s \cdot \gamma \cdot a^2}{N \cdot E_{bind}} = 1 - \frac{N_s \cdot \gamma \cdot a^2}{N \cdot E_{bind}} \sim 1 - \frac{a^3 \cdot \gamma}{d \cdot E_{bind}}$$
$$(1.4)$$

For Ag, where $E_{bind} \sim 160 \, \mathrm{kJ/mol}$, assuming 15% of surface atoms (see examples above), we obtain that $E \sim 0.76 \; E(bulk)$. That is, nanoparticles are less stable than the bulk.

The decreased stability of nanoparticles compared to the bulk has several consequences. The most immediate is that nanoparticles tend to aggregate to reduce their surface-to-volume ratio and thus the surface energy penalty, much like water droplets. Chemists call this process coagulation (which leads to precipitation), engineers may call it sintering. They are just different words to indicate the same concept. The different parlance used by different disciplines is a quite common occurrence in interdisciplinary fields such as nanotechnology. It may take some time and a few misunderstandings before an interdisciplinary team agrees on terminology and starts speaking the same language! The bottom line is that the sintering tendency can cause quite a few headaches. For example, chemists must protect nanoparticle in suspensions with surfactants to prevent precipitation. Supported nanoparticles, such as those used in heterogeneous catalysis, must be prevented from diffusing and aggregating, a quite complicated enterprise, especially in high temperature applications.

Coming back to our back-of-the-envelope calculation, we note that Eq. (1.4) shows that the depression of the cohesive energy of a nanoparticle goes with $1/d$. This is what we expect, since the fraction of surface atoms also goes as $1/d$, see Eq. (1.3). More rigorous calculations yield the same $1/d$ dependence on particle size. For example, Ref. [1] shows the melting temperature of a nanoparticle of radius r, $T_{melt}(r)$, to the melting temperature of the bulk, $T_{melt}(bulk)$, as follows:

$$T_{melt}(r) = T_{melt}(bulk) \left[1 - \frac{3}{\rho_s \Delta H r} \left(\gamma_s - \gamma_l \left(\frac{\rho_s}{\rho_l} \right)^{\frac{2}{3}} \right) \right], \quad (1.5)$$

where

γ_s = surface energy of the solid phase.
γ_l = surface energy of the liquid phase.
ρ_s = density of the solid phase.
ρ_l = density of the liquid phase.
ΔH = enthalpy of fusion.

Equation (1.5) has been successfully used in the past to fit the size dependence of the melting point of Au nanoparticles, as shown in

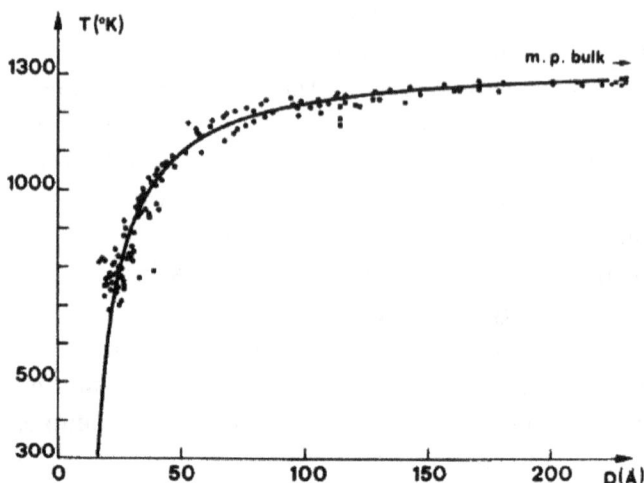

Figure 1.3:   Solid line: fit of Eq. (1.5) to experimental measurements of the melting point of Au nanoparticles as a function of their size. Adapted with permission from [2]. © American Physical Society.

Figure 1.3 [2]. The same work reports several phenomenological theories of nanoparticle melting, and it represents an excellent starting point for the study of this field.

Because of its relevance for applications such as welding, the *1/r* dependence of the melting point has been investigated by several groups. Figure 1.4 shows the melting point of In nanoparticles embedded into porous solids [3]. The *1/r* dependence is quite clear, and one may conclude that it is a universal law.

Well, not so fast. Melting is a quite complicated process. For starters, melting rarely involves the whole nanoparticle. Rather, the surface melts before the core. This can be explained by the lower coordination of the surface atoms compared to those in the core of the nanoparticle. Pioneering experiments and a theoretical model are reported in Ref. [4] for Sn nanoparticles, showing that the heat of fusion must be corrected for the particle size as follows:

$$\Delta H_{(particle)} = \Delta H_{(bulk)} \left(1 - \frac{t_0}{r} \right)^3, \qquad (1.6)$$

where t_0 is a fitting parameter.

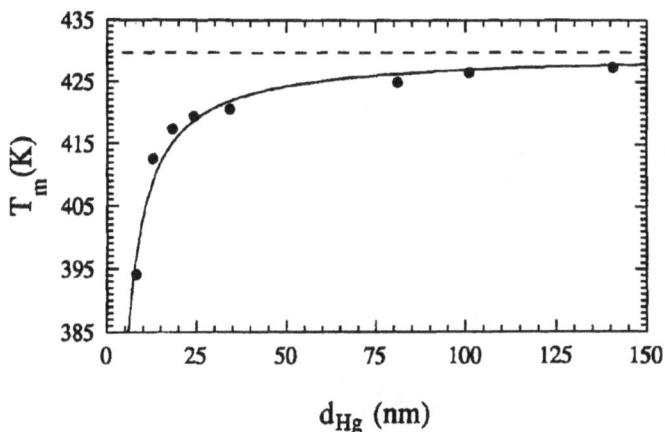

Figure 1.4: Dependence of the melting point of In nanoparticles embedded into porous materials. The mean pore size is termed d_{Hg} because it was determined by Hg porosimetry. Adapted with permission from [3]. © Elsevier Ltd.

Theoretical models also struggle to fit the experimental data when particles are not spherical. For example, In particles were deposited on a WSe_2 substrate and their melting point was measured as a function of size [5]. The nanoparticles had a triangular prismatic morphology, as shown in Figure 1.5 (left panel). The authors showed that the size dependence of the melting temperature (right panel) was much weaker than expected by theoretical models, most of which would yield a trend as shown by the dashed curve. The deviation from the theoretical expectations was attributed to the presence of a large fraction of low-index facets in triangular structures compared to spherical structures. Atoms in these low-index facets are quite stable, and therefore a comparatively higher energy is required for their melting. In case someone wonders why many studies are carried out on carbon (in this case, WSe_2), the answer is that these substrates typically interact very weakly with supported nanoparticles. Strong particle-substrate interactions would complicate the understanding of an already quite complex problem.

Now, one could think that the reduction of the melting point of nanoparticles is only a nuisance. It is in some fields at least; think catalysts, thin films made of isolated nanoparticles, or nanocomposites.

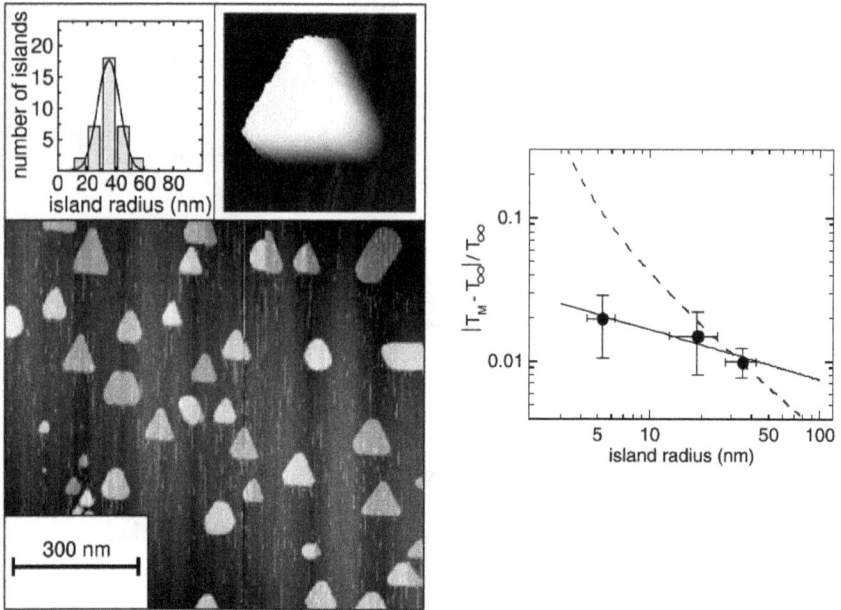

Figure 1.5:   Left panel: Scanning tunneling microscope images of In nanoparticles supported on WSe$_2$. Right panel: Black dots, experimental melting temperature. Dashed line: dependence of melting temperature on particle size as predicted by theoretical models assuming a spherical shape and a molten surface layer. Solid line: empirical fit to the data. Adapted with permission from [5]. © American Physical Society.

For such systems, the depression of the melting point may represent the difference between success and failure. However, in other fields, such as welding, or ceramic sintering, depression of the melting point is a boon. Let us look at Figure 1.6, which reports the sintering temperature of ceramics as a function of the mean particle size [6]. The data shows that reduction of the particle size can reduce the sintering temperature by 200–300°C. Power dissipation goes with the *fourth* power of the absolute temperature (Stefan-Boltzmann law), hence one can understand the relevance of nanoparticle research in the ceramic industry. A quite thorough review of the field of sintering of ceramic nanoparticles can be found in Ref. [7].

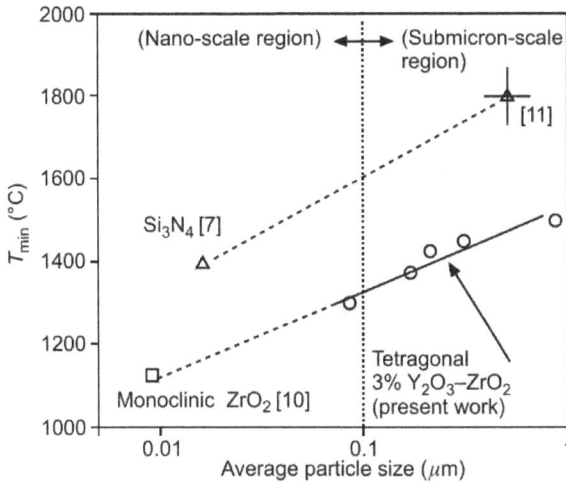

Figure 1.6:   Dependence of the sintering temperature on particle size for two ceramic systems. Adapted with permission from [6]. © Springer Nature.

1.4. Lattice contraction

Pioneering measurements of the lattice constant of nanoparticles were carried out by electron diffraction of nanoparticles of Ni and Cu deposited on amorphous carbon [8], whose results are reported in Figure 1.7.

A contraction of the lattice constant was observed, scaling as $1/d$, where d is the nanoparticle diameter. The contraction of the lattice constant is not unexpected, since it is a way of reducing surface area. The $1/d$ dependence is also in line with what we have seen so far for the melting temperature of nanoparticles. However, one must not forget that most solids are quite stiff. Hence, a lattice contraction carries a significant elastic energy penalty. A simplified yet quite effective theory has been reported by Qi et al. [9]. In this model, the lattice constant is assumed to change from its bulk value a_0 to a (contracted) value $a = a_0(1 - \varepsilon)$, where ε is the absolute value of lattice strain. The change in Helmholtz energy is calculated as the difference between the surface energy occupied by a surface atom with the bulk lattice constant $(\gamma\, a_0^2)$, the surface energy of the same atom with a contracted lattice constant $(\gamma\, a^2)$ and by an additional contribution

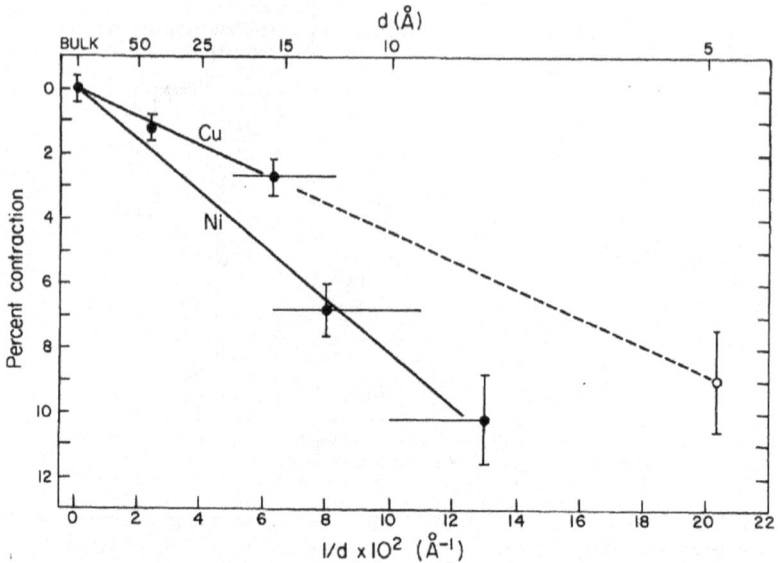

Figure 1.7: Lattice contraction of Ni, Cu nanoparticles. Adapted with permission from [8]. © American Physical Society.

due to the elastic energy of the strained lattice ($\frac{1}{2} G a_0^3 \varepsilon^2$). The elastic energy contains the rigidity modulus (G), which is the three-dimensional analog of the elastic constant of a spring. Since G has units of N/m^2, we must multiply by the volume of the atom to obtain energy units. Note: multiplying by a^3 or a_0^3 does not make much difference. Since ϵ is small, the elastic term retains only the term with the smallest power of ϵ. We also note that the surface contribution has to be multiplied by the number of surface atoms (N_s), while the elastic energy term is multiplied by the total number of atoms in the nanoparticle (N).

Hence:

$$\Delta F = N_s \gamma (a_0^2 (1 - \epsilon)^2 - a_0^2) + N 1/2 G \epsilon^2 a_0^3 \qquad (1.7)$$

The strain is determined by minimizing the Helmholtz energy:

$$\frac{d\Delta F}{d\epsilon} = 0, \quad \text{that is,} \quad \epsilon = \frac{1}{1 + \frac{G a_0}{2\gamma} \frac{N}{N_s}}, \qquad (1.8)$$

which once again yields the familiar $1/r$ relationship via the N/N_s term. (r = particle radius). Recent work has also shown that in small nanoparticles changes in the lattice can be more complex than a mere contraction. For example, the work by Fu *et al.* on Sn and Bi deposited electrochemically shows a lattice contraction with decreasing particle size, as well as a lattice distorsion [10]. Size effects can also lead to an altogether different lattice symmetry. For example, CdS has a comparatively open wurtzite structure for nanoparticle sizes >8–10 nm, but it has a more compact face-centered cubic structure below about 8 nm [11].

Recent reviews of experimental data, complemented by additional measurements, show that surface interactions and contaminants affect the lattice constant probably more strongly than thermodynamics alone. For example, Henry *et al.* [12] noticed an expansion of the lattice constant of Pd nanoparticles deposited on MgO. This expansion was attributed to strong interaction of the metal with the substrate lattice, which has a larger lattice constant. However, Lamber *et al.* [13] noticed a $1/r$ contraction on (non-supported) Pd nanoparticles produced by an inert gas evaporation technique. Szczerba *et al.* [14] carried out a very accurate analysis of citrate-capped Au nanoparticles. They noticed a contraction of the lattice constant which was 7–10 times smaller than that reported for supported nanoparticles of the same size. The lesson which these examples teach us is quite simple, yet extremely relevant. In nanotechnology, one must be very careful in separating *intrinsic* from *extrinsic* effects.

1.5. Reduction of thermal conductivity (Knudsen effect)

Another effect that takes place at the nano-scale and does not need to invoke new physics is the reduction of thermal conductivity that is encountered in foams with small pores. In foams, the thermal conductivity λ can be written as the sum of three different contributions:

$$\lambda = \lambda_{solid} + \lambda_{gas} + \lambda_{rad}, \tag{1.9}$$

Where λ_{solid} indicates the thermal conductivity through the solid phase, λ_{gas} the thermal conductivity through the gas contained in the foam pores, and λ_{rad} the radiant (infrared) contribution. Details of the theory of thermal conductivity can be found in text-books such as the one on foams by Gibson and Ashby [15], the one on cryogenic thermal insulation by Kaganer [16] and in aerogel liter-ature, for example by the group of Reichenauer [17, 22]. We will now look in detail at each of the terms in Eq. (1.9). The λ_{rad} term comes from the Stefan-Boltzmann law and accounts for less than 10% of the total thermal conductivity at room temperature. It is also indepen-dent of the structure of the material; hence we will neglect it in the discussion. The λ_{solid} term is more interesting. In conventional foams such as the polyurethane foams that can be found in most furniture, the solid contribution is expressed as [15]:

$$\lambda_{solid} = \lambda(bulk)\frac{\rho}{\rho(bulk)} \times \psi, \qquad (1.10)$$

where $\lambda(bulk)$ indicates the thermal conductivity of the bulk material (i.e., the polymer or whatever other material the foam is made of), ρ is the density of the foam, $\rho(bulk)$ is the bulk density, and ψ is a con-stant which depends on the geometry of the foam. In the most sim-ple cases, $\psi = 1/3$. This fraction is easy to understand: if we deliver energy to a particle that is connected in 3 dimensions, the energy will have equal probability of going in either direction. Now let us make a foam out of nanoparticles. Nanoparticles are messy objects, especially at their surfaces. Hence, the contact points between nanoparticles will have lots of defects, amorphous regions and what not. Therefore the solid thermal conductivity of nano-foams will be smaller than that predicted by Eq. (1.10). The largest nano-related effect, however, is on λ_{gas}. Kinetic theory of gases applied to the case of energy transfer between two walls separated by a distance δ is complex. A helpful parameter is the Knudsen number K_n:

$$K_n = \frac{L}{\delta}, \qquad (1.11)$$

where L is the mean free path of air molecules, which is $\sim$66 nm in dry air at standard temperature and pressure (STP) conditions [18]. When $K_n \ll 1$, the mean free path is much smaller than the dis-tance between the walls. Air molecules bounce into each other before hitting the walls. These gas phase collisions are independent on gas pressure and are dominated by the viscosity of the fluid. For these systems with large pores standard kinetic gas theory applies:

$$\lambda_{gas} = \epsilon \mu c_v,$$
$$\epsilon = \frac{9\gamma - 5}{4}, \tag{1.12}$$
$$\gamma = \frac{c_p}{c_v},$$

where μ is the viscosity of the gas, c_p is the specific heat at constant pressure and c_v is the specific heat at constant volume. Let us now look at Eq. (1.12) and the values reported in Table 1.1. We note that the thermal conductivity depends strongly on the mass of the gas for monoatomic gases, varying by about 30 times between He and Xe. The viscosity also depends on the gas, but not as strongly as the thermal conductivity. So, the only thing in Eq. (1.12) which could explain the thermal conductivity trend is c_v. Alas, kinetic theory states that $c_v = 3/2R$ for monoatomic gases. Yes, there may be some non-ideality correction, but we know that these corrections are not large in rare gases. So, what gives? The trick is to recognize that c_v in Eq. (1.12) needs to be provided in J/kg, not in J/mol as we usually do when dealing with kinetic theory. Then, the trends of Table 1.1 make sense! And they also explain why double-pane windows are filled with Ar and not with He. Xe would be of course the ideal choice for window insulation, but there are cost issues related to Xe. Windows and c_v digressions

Table 1.1: Thermal conductivities and viscosities of selected gases.

Gas	$\lambda(\mathrm{mW \cdot m^{-1} \cdot K^{-1}})$[a]	$\mu(10^{-5}\,\mathrm{Pa \cdot s})$[b]
He	155.7	1.96
Ne	49.4	3.36
Ar	17.7	2.23
Kr	9.5	
Xe	5.5	2.51
Air	26.4	1.82

[a]STP, [b]$P = 1$ atm, $T = 20°$C. See also [19].

aside, we note that in foams $\lambda_{gas} \sim \lambda_{air} \sim 0.026\,\mathrm{W} \cdot \mathrm{m}^{-1} \cdot \mathrm{K}^{-1}$ when $K_n \ll 1$.

When $K_n \gg 1$, the mean free path is larger than the distance between the walls. Molecules move from one surface to the other without colliding with each other. They behave like bullets emitted from one surface and hitting the other surface. In this case, kinetic theory yields the following expression for the thermal conductivity:

$$\lambda_{gas} = \frac{\gamma+1}{\gamma-1}\sqrt{\frac{R}{8\pi}\frac{p}{MT}}\alpha, \tag{1.13}$$

where M is the mass of the gas, p its pressure, and α is the accommodation coefficient, a parameter that expresses the efficiency of the energy transfer between the colliding molecule and the surface. Thorough analysis of the accommodation coefficient has been carried out by Saxena *et al.* [20] and, more recently, by Manson [21] and an often-used expression is:

$$\alpha = 1 - \left(\frac{M-m}{M+m}\right)^2, \tag{1.14}$$

where M is the mass of the projectile and m is the mass of the surface (target) molecules. From Eq. (1.14) we see that $\alpha \sim 1$ when the mass of the projectile is comparable to the mass of the surface, and that $\alpha \sim 0$ when the mass of the projectile is small compared to the mass of the surface. Now, many porous systems have pores on the order of 10–20 nm. This is a bit of a headache, since in this case $K_n \sim 1$ and there is no hand-waving explanation. For this intermediate case, the thermal conductivity is given by [16, 22]:

$$\lambda_{gas} = \frac{\lambda_0}{1+2\beta K_n},$$

$$\beta = \frac{2\epsilon}{\gamma+1}\frac{2-\alpha}{\alpha}, \tag{1.15}$$

where λ_0 is the thermal conductivity of the pore-filling gas. Plotting λ_{gas} as a function of the various parameters is an instructive and highly recommended exercise!

The group of Reichenauer [22] compared the thermal conductivity as a function of pressure for a series of porous materials. The data

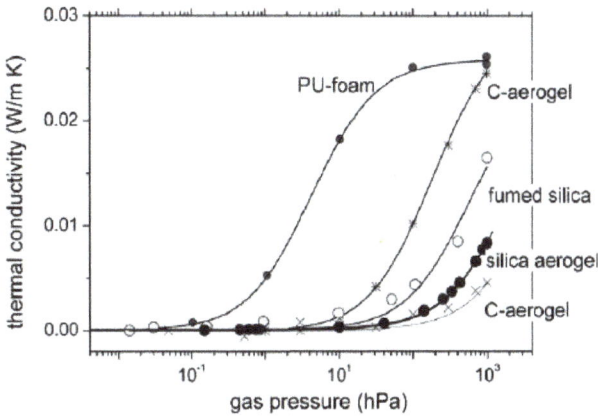

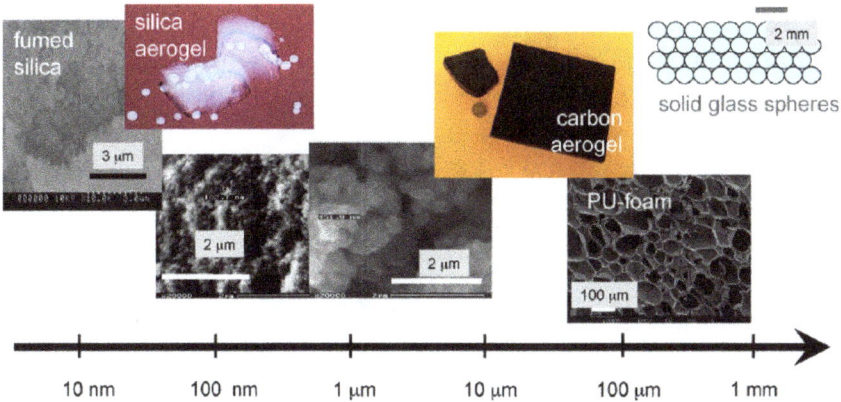

Figure 1.8: Thermal conductivity as a function of air pressure for a series of porous materials. Adapted with permission from [22]. © Elsevier.

was fitted using Eq. (1.15), and the results are reported in Figure 1.8. The agreement between theory and experiment is remarkable, and surprisingly, is valid also when K_n is not on the order of 1, that is, outside the range for which Eq. (1.15) was derived.

1.6. Oversolubility

Nanoporous materials have a high surface area, and are therefore candidates for the development of absorbents. This is nothing new or surprising, of course. What is surprising is a higher-than expected solubility of gases in solvents confined in small pores. This effect is called oversolubility and was first observed in membrane reactors [23]. Oversolubility is a complex phenomenon. An explanation was provided by Micahon *et al.* [24] The authors filled porous materials (commercial alumina and silica, often used as catalyst support) with an organic solvent such as CCl_4, and measured the solubility of gases such as H_2, CH_4 and C_2H_6. The authors noted that the solubility of the gases was comparable to that in bulk organic liquid when the pores were completely filled with liquid. However, the solubility increased by 2–10 times when the pores were only partially filled with the solvent. The effect did not depend on the solvent [25], and data for ethanol is reported in Figure 1.9. The authors interpreted the oversolubility as deriving from a surface effect. The surfaces of liquids are "fuzzy", and consist of a rarefied layer on top of the liquid, as shown in Figure 1.10. In this rarefied layer solvent and

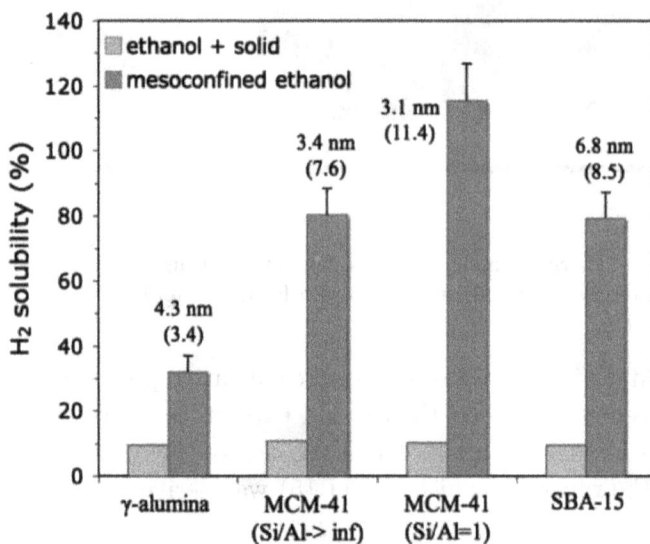

Figure 1.9: Oversolubility of H_2 in ethanol. Porous solids with the mean pore sizes indicated in the figures were completely (lower bar) and partially filled (higher bar) with ethanol. The numbers in parentheses indicate the increase in solubility induced by partial filling. Values higher than 100% are attributed to interactions with the pore walls. Adapted with permission from [25]. © Elsevier.

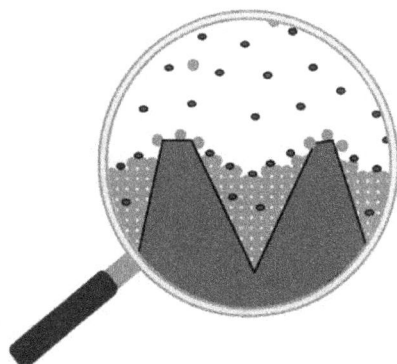

Figure 1.10: Schematic representation of the rarefied solvent layer inside pores partially filled with an organic liquid. Adapted with permission from [24]. © Wiley-VCH.

solute are both in the gas phase, where solubility is increased. The high surface area of the nanoporous matrix ensures a large capacity of these sorbents. Oversolubility is a promising way getting rid of pollutants, the most prominent of which is CO_2. CO_2 is captured industrially by bubbling gas into liquid amines. While cost-effective, liquid capture is far from ideal, and solid state systems would be preferable to liquids. Hence, CO_2 oversolubility is being actively investigated.

Recent work has shown that oversolubility can also occur in porous systems completely filled with an organic solvent. For example, in Ref. [26] it was shown that CO_2 solubility was increased by up to 30% in nanoporous systems filled with organic solvents. The effect was attributed to the formation of an ordered solvent structure inside the pores. The CO_2 solute could adhere to the pore walls, but could also infiltrate in the spaces between the solvent molecules, as shown in Figure 1.11. Formation of an ordered solvent layer with internal cavities was confirmed by Monte Carlo simulations which showed formation of cavities in methanol capable of hosting CO_2 molecules [27]. Hence, once again, let us not forget the guiding principle of this chapter. Effects do occur at the nanoscale that are interesting and technologically very relevant. However, in many cases no new physics is needed to explain these effects.

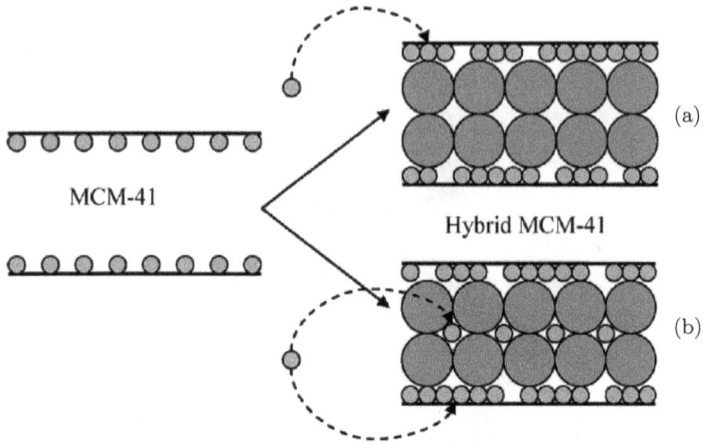

Figure 1.11: Schematic representation of the oversolubility mechanism of CO_2 in pores completely filled with an organic solvent. CO_2 (smaller circles) adsorbs to the pore walls and between the spaces of the ordered structure formed by the solvent (larger circles) inside the pores. Adapted with permission from [26]. © American Chemical Society.

1.7. Quantum dots

Quantum dots are a prime example of what happens when the size of a particle is reduced. Before tackling quantum dots, however, we need to talk about excitons. In general, when we think of a material absorbing a photon, we visualize an electron going from the valence to the conduction band, then drifting away and leaving a positive charge (hole) behind. However, this is not always the case.

One can think of the electron-hole pair as forming a hydrogen atom. The electron will orbit around the hole as shown in Figure 1.12. The Coulombic interaction between the mobile electron and the fixed hole has the same form than in the hydrogen atom:

$$E_{exciton} = \frac{-e^2}{\epsilon r},$$

where e is the charge of the electron and ϵ is the dielectric constant of the material.

The energy levels look as in Fig. 1.12. The important thing to notice is that in the bulk, excitons have an energy **smaller** than the band gap. The other important thing to notice is that excitons have

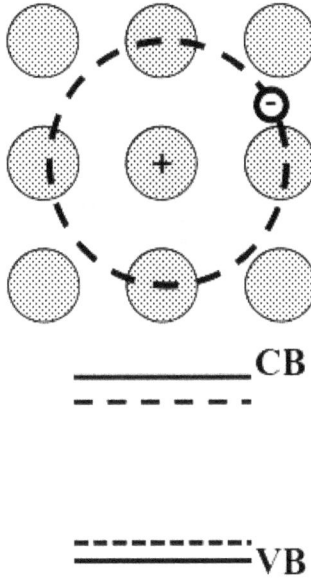

Figure 1.12: Top: depiction of an exciton. Bottom: corresponding energetic levels in the bulk. Dashed lines indicate the excitonic levels.

a radius. This radius is the Bohr radius of the hydrogen atom (a_B), increased by the dielectric constant: $a_{exciton} \propto a_B \times \epsilon$. The larger radius (compared to the hydrogen atom) is due to the screening of the electric field by the dielectric constant of the material.

Let us now consider the same situation (electron and hole bound in an exciton) but inside a small particle, as shown in Fig. 1.13. In the Hamiltonian for such a system, there will be two contributions to the energy. One contribution will be the electrostatic energy $E_{Exciton} = \frac{-e^2}{\epsilon r}$ and another contribution will come from quantum confinement, since the electron can also be considered to be confined within a quantum well. The quantum confinement energy for an infinite well is $E_{QC} = \frac{\hbar^2}{mr^2}$ [28]. The quantum confinement energy has a $1/r^2$ dependence and will therefore prevail over the electrostatic energy at small sizes. When the quantum confinement energy prevails, the energetic levels of the system behave like those of a particle in a box:

$$E_n = \frac{(\hbar \pi n)^2}{2mr^2}, \qquad (1.16)$$

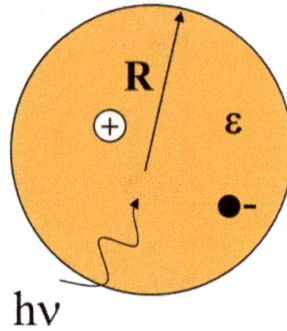

Figure 1.13: Creation of an exciton inside a nanoparticle of radius r with a dielectric constant ϵ.

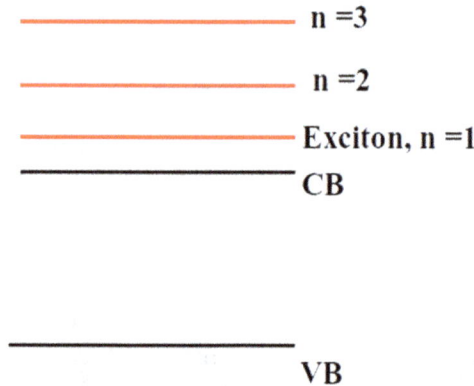

Figure 1.14: Schematic representation of the excitonic energetic levels in a quantum-confined semiconductor nanoparticle.

where r is the particle radius, m the mass of the electron and n the quantum number (integer). The rule of thumb for quantum confinement is that the radius of the particle must be smaller than the radius of the exciton. When this is the case, the excitonic levels are placed as Figure 1.14. One should note that in nanoparticles the energies of the excitons are HIGHER than the band gap, and not LOWER as in the bulk. Also, the energy of the levels and of the band gap goes as $1/r^2$. The dependence of absorption and emission spectra of quantum dots as a function of their size has been reported by a lot of researchers.

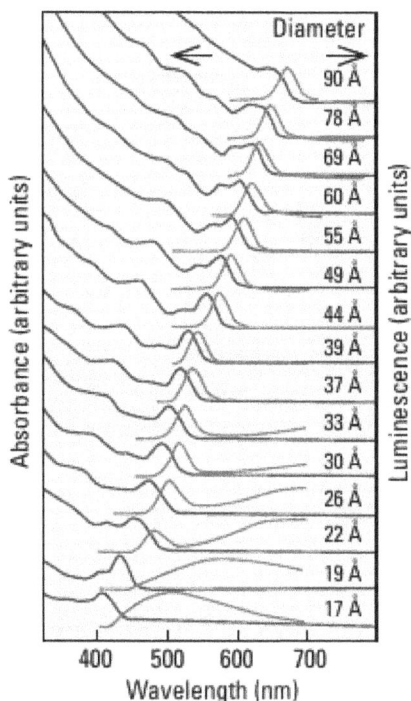

Figure 1.15: Dependence on size of the optical properties of CdSe quantum dots. Dark grey trace: absorption. Light grey trace: emission. Adapted with permission from [29]. © American Chemical Society.

The reader will easily find on the internet pretty pictures of quantum dots covering the colors of the rainbow [29, 30]. In Figure 1.15, absorption and emission spectra are reported as a function of size. The shift of absorption and emission towards higher energies (shorter wavelengths) with decreasing size is evident. One also notes that for sizes below about 3.3 nm a broad peak appears at low energies. This peak is due to defects, which, due to the small particle size, are surface defects. In the simplest case, defects introduce levels within the band gap and lower the emission energy. A schematic representation of the lowering of the emission energy is shown in Figure 1.16. However, defects can also capture electrons. If the defects are shallow, the electrons will be released after typically a few milliseconds. However, if the traps are deep, electrons may remain trapped indefinitely. In

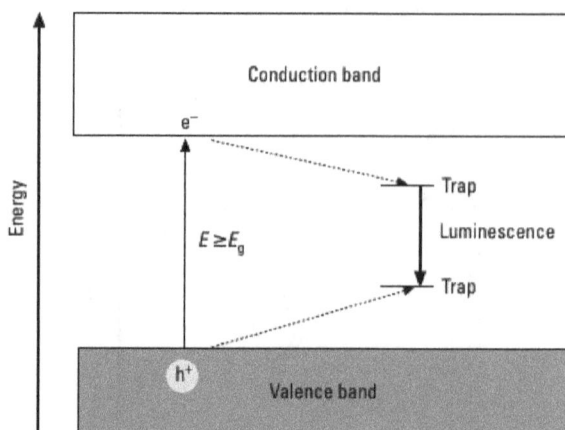

Figure 1.16: Schematic representation of the introduction of energetic levels within the band gap by defects. Adapted with permission from [29]. © American Chemical Society.

the early days, quantum dots had a lot of defects, and their quantum yield (i.e., emission efficiency) was ∼1%, sometimes less. These days, quantum yields > ∼10% appear to be the norm [18]. Before closing, two issues must be emphasized. The first is that the study of defects is a field in and of itself. Too often, incorrect attributions of emission peaks are reported in the literature. Before attributing peaks to this or that defect, one should consult reputable sources, which are not necessarily the ones that are most often cited [31]. In addition, one should be warned that often times reports appear in the literature where the term "quantum confinement" is abused, and/or luminescence is reported as arising from nano-objects, which is instead due to defects. In this context, a search for "carbon quantum dots" and "luminescence of Au" may be illuminating.

References

1. See, for example, P. Letellier, A. Mayaffre, and M. Turmine. *Physical Review B* **76**, 045428 (2007).
2. P. Buffat and J.-P. Borel, *Physical Review A* **13**, 2287 (1976).
3. K. M. Unruh, J. F. Sheehan, "Nanostructured Materials" Vol. 3, pp. 425 431 (1993)

4. S. L. Lai, J. Y. Guo, V. Petrova, G. Ramanath, L. H. Allen, *Phys. Rev. Lett.* **77**, 99 (1976).
5. M. Dippel, A. Maier, V. Gimple, H. Wider, W. E. Evenson, R. L. Rasera, and G. Schatz, *Phys. Rev. Lett.* **87**, 095505 (2001).
6. Journal of Materials Science Letters 17 (1998) 467 ± 469 DEAN-MO LIU.
7. Z. Z. Fang and H. Wang, International Materials Reviews 2008 Vol. 53, 326.
8. G. Apai, J. F. Hamilton, J. Stohr, A. Thompson, *Phys. Rev. Lett.* **43**, 165 (1979).
9. W. H. Qi, M. P. Wang, Y. C. Su, *Journal of Materials Science Letters* **21**, 877–878 (2002).
10. X. F. Yu, X. Liu, K. Zhang, and Z. Q. Hu, *J. Phys.: Condens. Matter* **11**, 937 (1999).
11. J. Marquez-Marín, C. G. Torres-Castanedo, G. Torres-Delgado, M. A. Aguilar-Frutis, R. Castanedo-Perez, and O. Zelaya-Angel, 10.1016/j.spmi.2017.01.007. *Superlattices and Microstructures* **102**, 442e450 (2017).
12. S. Giorgio, C. R. Henry, C. Chapon, J. M. Penisson, *J. Cryst. Growth.* **100**, 254 (1990).
13. R. Lamber, S. Wetjen, N. I. Jaeger, *Phys. Rev. B* **51**, 10968 (1995).
14. W. Szczerba, H. Riesemeier, A. F. Thunemann, *Analytical and Bioanalytical Chemistry* **398**, 1967–1972 (2010).
15. Ashby and Gibson, Cellular Solids, 2nd Ed., Cambridge Solid State Science Series, 1997.
16. M. G. Kaganer, "Thermal insulation in cryogenic engineering", Israel Program for Scientific Translations, Jerusalem, 1969.
17. L. Weigold, D. P. Mohite, S. Mahadik-Khanolkar, N. Leventis, G. Reichenauer, *Journal of Non-Crystalline Solids* **368**, 105–111 (2013).
18. S. G. Jennings, *Journal of Aerosol Science* **19**, 159–166 (1988).
19. https://ws680.nist.gov/publication/get_pdf.cfm?pub_id=907540
20. S. C. Saxena, R. K. Joshi, "Thermal accommodation and adsorption coefficient of gases", in: C.Y. Ho (Ed.), CINDAS Data Series on Material Properties, Hemisphere Publishing Co., New York, 1989.
21. G. Fan, J. R. Manson, *Chemical Physics* **370** 175–179 (2010); A. Muis and J. R. Manson, *Surf. Sci.* **486**, 82–94 (2001).
22. G. Reichenauer, U. Heinemann, H.-P. Ebert. Colloids and Surfaces A: Physicochem. *Eng. Aspects* **300**, 204–210 (2007).
23. J. Peureux, M. Torres, H. Mozzanega, A. Giroir-Fendler, J. A. Dalmon, *Catal. Today* **25**, 409–415 (1995).
24. S. Miachon, V. V. Syakaev, A. Rakhmatullin, Marc Pera-Titus, Stefano Caldarelli, and Jean-Alain Dalmon, *Chem. Phys. Chem.* **9**, 78–82 (2008).
25. V. Rakotovao, R. Ammar, S. Miachon, Marc Pera-Titus, *Chemical Physics Letters* **485**, 299–303 (2010).
26. N. Linh Ho, J. Perez-Pellitero, F. Porcheron, and R. J.-M. Pellenq, *J. Phys. Chem. C* **116**, 3600–3607 (2012).
27. E. Sánchez-González, P. G. M. Mileo, J. Raziel Álvarez, E. González-Zamora, G. Maurin, and I. A. Ibarra, *Dalton Trans.* **46**, 15208 (2017).
28. L. E. Brus, *J. Chem. Phys.* **79**, 5566 (1983).

29. C. J. Murhpy, Optical sensing with quantum dots, *Analytical Chemistry* 74, 19, 520 A–526 A(2002).

30. B. O. Dabbousi, J. Rodriguez-Viejo, F. V. Mikulec, J. R. Heine, H. Mattoussi, R. Ober, K. F. Jensen, and M. G. Bawendi, *J. Phys. Chem. B* **101**, 9463–9475 (1997).

31. M. A. Reshchikov, H. Morkoc, B. Nemeth, J. Nause, J. Xie, B. Hertog, A. Osinsky, *Physica B* 401–402, 358–361 (2007).

Chapter 2

Characterization Techniques

It may appear a bit surprising to discuss characterization techniques at the beginning of an introductory textbook in nanotechnology. However, when one talks about nanoparticles, characterization techniques play an important role. Hence, the earlier these techniques are discussed, the better.

2.1. Waves, direct and reciprocal space

Diffraction techniques are one of the workhorses of materials science and nanotechnology. Before getting into them, it is a good idea to refresh wave propagation and introduce the concept of lattice, direct space and reciprocal space.

2.1.1. *Waves and electromagnetism: A primer and a refresher*

With a few exceptions, waves are defined by the equation:

$$\frac{\partial^2 f}{\partial z^2} = \frac{1}{v^2} \frac{\partial^2 f}{\partial t^2},$$

(2.1)

where f is a suitable, twice differentiable function.

A solution of (2.1) is:

$$f(z,t) = A \cdot \cos(kx - \omega t + \delta). \tag{2.2}$$

Substitute (2.2) into (2.1) if you are not convinced. In Eq. (2.2), k represents the periodicity of the wave in space and ω the periodicity in time. Let us see more in detail what this means. We know that waves have a wavelength of λ. This means that $f(z + \lambda) = f(z)$. Assume $t = 0$, $\delta = 0$ in Eq. (2.2), and this means $\cos((k + \lambda) \cdot z) = \cos(k \cdot z)$. This means

$$k \cdot \lambda = 2\pi, \text{ that is, } k = \frac{2\pi}{\lambda}. \tag{2.3}$$

Similarly, knowing that waves repeat with period T, we get $f(t+T) = f(t)$, that is, $\cos((t+T) \cdot \omega) = \cos(t \cdot \omega)$, or:

$$T \cdot \omega = 2\pi, \text{ that is, } \omega = \frac{2 \cdot \pi}{T}. \tag{2.4}$$

We also notice that the velocity of propagation of a wave is

$$v = \frac{\omega}{k} = \frac{\lambda}{T}. \tag{2.5}$$

Equation (2.5) is the "normal" dispersion relation. We will see when we discuss plasmons that other dispersion relations are possible, and that they lead to quite interesting results. Now, $f(z,t) = Ae^{i(kz-\omega t)}$ is the general solution of Eq. (2.1), where A is a complex number. Since we are interested only in the <u>real</u> part of the solution, we will do the math using the imaginary solution, but consider only the real part $f(z,t) = \mathcal{R}(Ae^{i(kz-\omega t)})$ when discussing the physics of the problem.

Let us now turn to Maxwell's Equations:

$$\nabla \cdot E = \frac{\rho}{\epsilon_0}. \qquad\qquad \nabla \cdot B = 0.$$

$$\nabla \times E = \frac{-\partial B}{\partial t}. \quad \nabla \times B = \mu_0 \cdot J + \mu_0\epsilon_0\frac{\partial E}{\partial t} \tag{2.6}$$

From these equations we can derive a wave equation:

$$\nabla \times (\nabla \times E) = \nabla(\nabla \cdot E) - \nabla^2 E = \nabla \times \left(\frac{-\partial B}{\partial t}\right)$$

$$= \frac{-\partial}{\partial t}(\mu_o \nabla \times J) - \mu_0 \epsilon_0 \frac{\partial}{\partial t}\left(\frac{\partial E}{\partial t}\right). \tag{2.7}$$

When sufficiently far away from charges and currents (i.e., in free space), $\rho = 0$ and $J = 0$, and we are left with:

$$-\nabla^2 E = \epsilon_0 \mu_0 \frac{\partial^2 E}{\partial t^2}, \tag{2.8}$$

which in one dimension becomes

$$\frac{\partial^2 E}{\partial z^2} = -\epsilon_0 \mu_0 \frac{\partial^2 E}{\partial t^2}, \tag{2.9}$$

which is indeed a wave equation.

The solution of Eq. (2.9) has the form:

$$E(z,t) = E_0 e^{i(kz-\omega t)}$$
$$B(z,t) = B_0 e^{i(kz-\omega t)}. \tag{2.10}$$

We note that we have TWO waves, one for the magnetic field, one for the electric field! Also, the waves are transverse. The propagation velocity in vacuum is $c = \frac{1}{\sqrt{\epsilon_0 \mu_0}}$ and in dielectrics:

$v = \frac{1}{\sqrt{\epsilon \mu}} = \frac{c}{n}$, where $n = \sqrt{\frac{\epsilon \mu}{\epsilon_0 \mu_0}}$. Since in most cases $\mu \cong \mu_0$, $n = \sqrt{\frac{\epsilon}{\epsilon_0}} = \sqrt{\epsilon_r}$, and ϵ_r is the dielectric constant of the material.

We note that ϵ_r can be real or complex. In case it is complex, for example in metals, also k, the wave vector, is complex:

$$k = k_0 + ik', \tag{2.11}$$

which means that the wave is attenuating when penetrating inside the material. To prove this, substitute Eq. (2.11) into Eq. (2.10).

Limiting the discussion to the electric field:

$$E(z,t) = E_0 e^{i(kz-\omega t)} = E_0 e^{-k'z} e^{i(k_0 z - \omega t)}, \qquad (2.12)$$

which yields an exponential decay in the direction of propagation of the wave. We will see more about wave propagation and absorption in Chapter 4.

2.1.2. *Direct lattice*

A solid is characterized by a periodic distribution of atoms or, better, of elementary units. We express periodicity by stating that units are on a lattice of points. We call the lattice a Bravais lattice and the points on the lattice are given by:

$$\mathbf{R} = \sum_{i=1}^{3} n_i \mathbf{a_i},$$

where n_i is an integer and $\mathbf{a_i}$ are three vectors, not in the same plane, which are called primitive vectors.

One can find on the Internet lots of examples of primitive vectors for two- and three-dimensional lattices, an excellent source being from the University of Cambridge.[1] For this reason, the representation of lattices will be omitted altogether.

2.1.3. *Reciprocal lattice*

Assume that a wave propagates through a lattice. The wave is described by a function of the type $e^{i\mathbf{k}\cdot\mathbf{r}}$ (note the vector notation for wave vector and position). There will be some wavelengths (i.e., values of $\mathbf{k}$) for which the wave has the same periodicity as the lattice, as shown in Figure 2.1.

The periodicity condition of Figure 2.1 is expressed by: $e^{i\mathbf{k}\cdot(\mathbf{r}+\mathbf{R})} = e^{i\mathbf{k}\cdot\mathbf{r}}$, or, equivalently, by:

$$e^{i\mathbf{k}\cdot\mathbf{R}} = 1. \qquad (2.13)$$

It turns out that the vectors $\mathbf{k}$ that fulfill the periodicity condition in Eq. (2.13) also span a lattice. This lattice is called reciprocal lattice

---

[1]https://www.doitpoms.ac.uk/tlplib/miller_indices/index.php

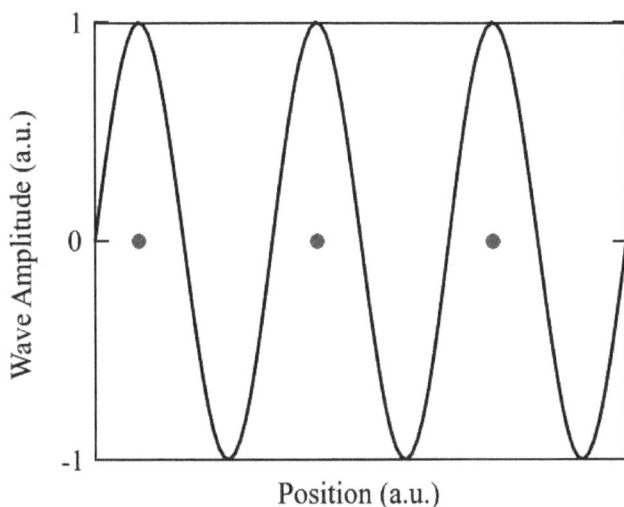

Figure 2.1: Example of a wave with the same periodicity of an underlying lattice, indicated by points.

and its primitive vectors are:

$$\mathbf{b_1} = 2\pi \frac{\mathbf{a_2} \times \mathbf{a_3}}{\mathbf{a_1} \cdot (\mathbf{a_2} \times \mathbf{a_3})},$$

$$\mathbf{b_2} = 2\pi \frac{\mathbf{a_3} \times \mathbf{a_1}}{\mathbf{a_1} \cdot (\mathbf{a_2} \times \mathbf{a_3})},$$

$$\mathbf{b_3} = 2\pi \frac{\mathbf{a_1} \times \mathbf{a_2}}{\mathbf{a_1} \cdot (\mathbf{a_2} \times \mathbf{a_3})}.$$

2.1.4. *Putting it together: reciprocal space, lattice planes and Miller indices*

If you look at a solid, it easy to identify planes; see for example Figure 2.2. Planes play an important role in diffraction. Let us see why.

From linear algebra, we should remember that a family of parallel planes can be identified by a vector that is perpendicular to one plane of that family. This definition is a close relative of the reciprocal lattice, Eq. (2.13). In fact, $e^{i\mathbf{k}\cdot\mathbf{R}} = 1$ means $\mathbf{k} \cdot \mathbf{R} = n\pi$. For $n = 0$: $\mathbf{k} \cdot \mathbf{R} = 0$, which is the geometric definition of perpendicularity. So, one vector in reciprocal space identifies a family of planes in the solid, parallel to each other. It gets better. A reciprocal lattice vector is given by: $\mathbf{k} = h\mathbf{b_1} + k\mathbf{b_2} + l\mathbf{b_3}$, h, k, l all integers. This vector identifies planes in real space that are perpendicular to it. The integers h, k, l are the Miller indices of the plane, and some examples are shown in Figure 2.3.

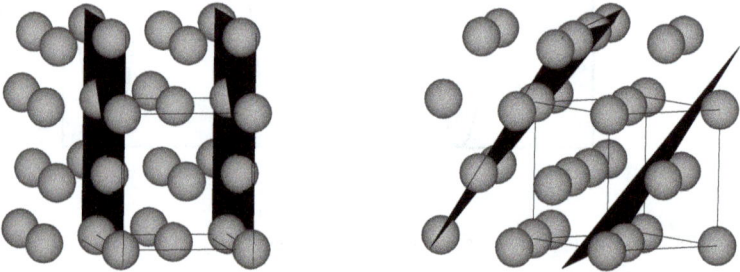

Figure 2.2:   Examples of planes in a fcc lattice.

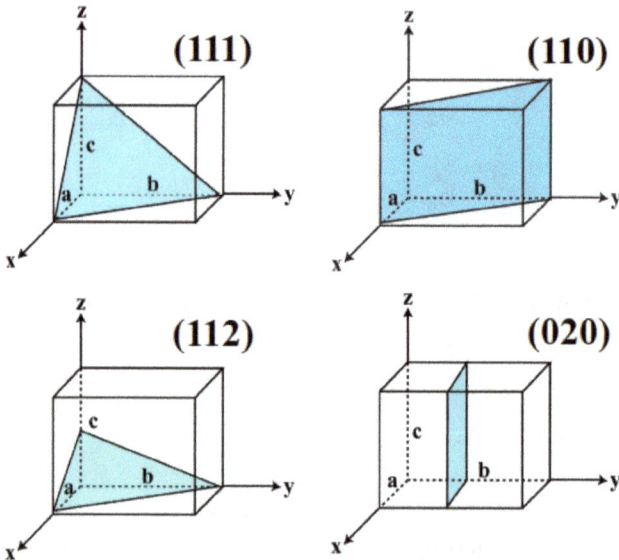

Figure 2.3:   Examples of planes defined by the Miller indices reported at the bottom of each cube. Images courtesy of JEOL, Inc.

2.1.5. *Bragg diffraction*

In a crystal, we can identify planes. Bragg, in a simplified model, assumed that X-rays (or other radiation types) are reflected back by planes of atoms, as shown in Figure 2.4.

The path difference between the beam reflected by the top plane and the plane underneath it is equal to $dsin(\theta)$; therefore the condition for constructive interference is

$$2d\sin(\theta) = n\lambda, \quad n \text{ integer} \tag{2.14}$$

which also means that the planes defined by the Miller indices are responsible for diffraction.

2.1.6. *Diffraction from nanoparticles*

To understand nano-effects on diffraction, we first have to look at the interaction of X-rays (or other radiation) with a crystal. For this, look at the sketch in Figure 2.5.

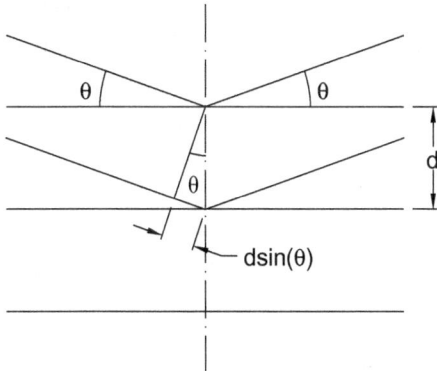

Figure 2.4: Schematic illustration of the Bragg model for diffraction.

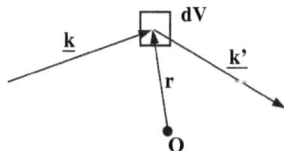

Figure 2.5: Schematic representation of the geometry used to calculate diffraction in solids.

An incident wave with wave vector **k** interacts with a volume element dV of a solid. The volume element is located at position **r**. The radiation is scattered with a wave vector **k′** (or, better, we look at the wave scattered with wave vector **k′**). Before we go on, note that **k** contains the magnitude of the wavevector (that is, the wavelength), but also the direction. That is why we use a vector notation for **k**. Also, for this discussion, we focus on elastic scattering; therefore $|k| = |k'|$. The interaction probability F of the radiation with dV depends (for X-rays at least) on the electron density n(**r**) and it can be written as:

$$F = \int dV \cdot n(\mathbf{r}) \cdot e^{i(\mathbf{k}-\mathbf{k}')\cdot\mathbf{r}}. \tag{2.15}$$

Equation (2.15) is for an infinite solid. For a finite number of atoms (say, M), the scattered intensity will be proportional to:

$$F = \sum_{i=1}^{M} e^{i(\mathbf{k}-\mathbf{k}')\cdot m\mathbf{a}}, \tag{2.16}$$

where we assume all M scatterers to be equal (so that n(**r**) is constant and can be ignored in the sum). Since Eq. (2.16) is a sum over a finite number of elements, it can be written as:

$$F = \frac{1 - e^{-iM\mathbf{a}\cdot\Delta k}}{1 - e^{-i a\cdot\Delta k}}. \tag{2.17}$$

In diffraction (neutron, X-ray, etc.) we measure not the amplitude by the square of the amplitude (which is the energy of a wave, let us not forget that). So:

$$F^2 = \frac{\sin^2(1/2M(\mathbf{a}\cdot\Delta\mathbf{k}))}{\sin^2(1/2(\mathbf{a}\cdot\Delta\mathbf{k}))}. \tag{2.18}$$

Call now $\mathbf{a}\cdot\Delta\mathbf{k} = \epsilon$ to simplify life. Equation (2.18) becomes $F^2 = \frac{\sin^2(1/2M\epsilon)}{\sin^2(1/2\epsilon)}$. This function is the diffraction spectrum from a small particle. Intuitively you can expect that this function will have some maxima and minima. The width of the maxima will be the width of the diffraction peaks. Since M is big, $\sin^2(1/2M\epsilon)$ is a rapidly varying function. In the denominator, $\sin^2(1/2\epsilon)$ is a slowly varying function which we can consider as constant. To determine the width of the peaks, one just needs the zeroes of $\sin^2(1/2M\epsilon)$. The first zero is for $\varepsilon = 0$, the second zero is for $M\varepsilon/2 = \pi$, that is, $\varepsilon = 2\pi/M$. That is, the diffraction peak has a width proportional to $1/M$. This means that the width of a diffraction peak will increase when the number of atoms in a particle is decreased.

An alternative description is reported in the caption of Figure 2.6. However, never forget that words help up to a point. Equation (2.18) is a more quantitative approach.

Here is an example from real life. TiO_2 nanoparticles were synthesized and deposited on a substrate [1]. The nanoparticles were quite small, around 3 nm. Correspondingly, in the X-Ray diffraction spectra reported in Figure 2.7, the lowest trace exhibits quite broad peaks. The peaks are at the angles expected for anatase TiO_2. Transmission electron microscopy (not presented here) had also shown that these particles were indeed crystalline, and not some amorphous stuff.

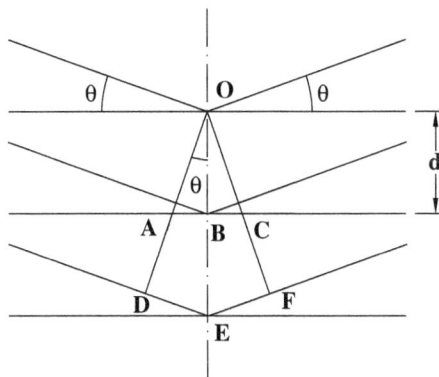

Figure 2.6: Alternative derivation of the broadening of diffraction peaks with decreasing size. When the diffraction conditions are met, the path difference between adjacent planes is exactly equal to an integer of λ. That is, $ABC = \lambda$, $DEF = 2\lambda$, etc. Assume now that the angle is a bit off the constructive diffraction conditions, so that $ABC = 1.1\lambda$, $DEF = 2.2\lambda$, etc. If we had only these two planes, we would see some signal at this angle. However, plane number 5 below the surface will have a path difference of $5.5\ \lambda$. Thus, light scattered from plane 5 will be $1/2\lambda$ out of phase with the surface plane and cancel out the reflection from it. Plane number 6 will cancel out plane number 2, etc, explaining why diffraction from infinite solids gives rise to narrow peaks. If the particle is small, there may not be plane number 6 to cancel out plane number 2. Hence, we will see signal even at angles which are off the diffraction condition.

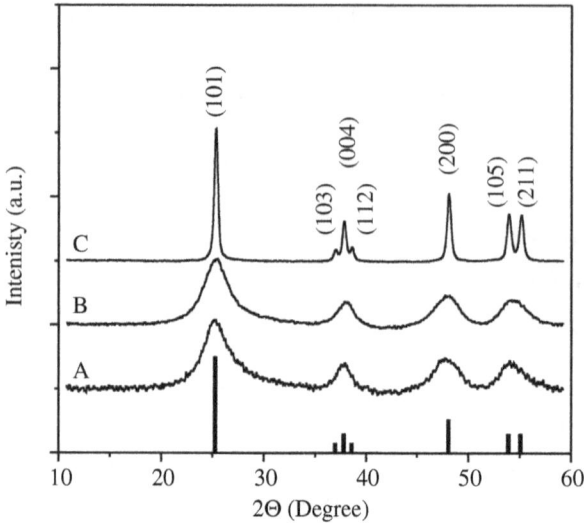

Figure 2.7:   XRD patterns of anatase nanoparticle films (A) as prepared, (B) UV C irradiated and (C) annealed at 500°C. The average grain size of (A), (B) and (C) is 3.2 nm, 3.2 nm and 20.6 nm, respectively. The stick pattern shows JCPDS (Joint Committee on Powder Diffraction Standards) reference card no. 21–1272 for single phase anatase. Adapted with permission from Wiley.

After annealing at 550°C, the particle size increased to about 25 nm, and the X-ray diffraction peaks became much narrower.

Another thing that can happen when analyzing nanoparticles is preferential growth in one direction [2]. In the experiment reported in Figure 2.8, Fe carbide particles were synthesized at room temperature and processed at 365°C for the indicated times. For short times, we note a spherical particle morphology and XRD shows a mixture of carbide and oxide phases. For increasing times, the particles assume a platelet-like configuration, and XRD shows that Fe_3C becomes the predominant phase. In addition, the peak at about 42 degrees becomes predominant because of the preferential growth of the base of the hexagonal prisms. The histograms below the XRD spectra indicate the intensities measured from standard bulk samples.

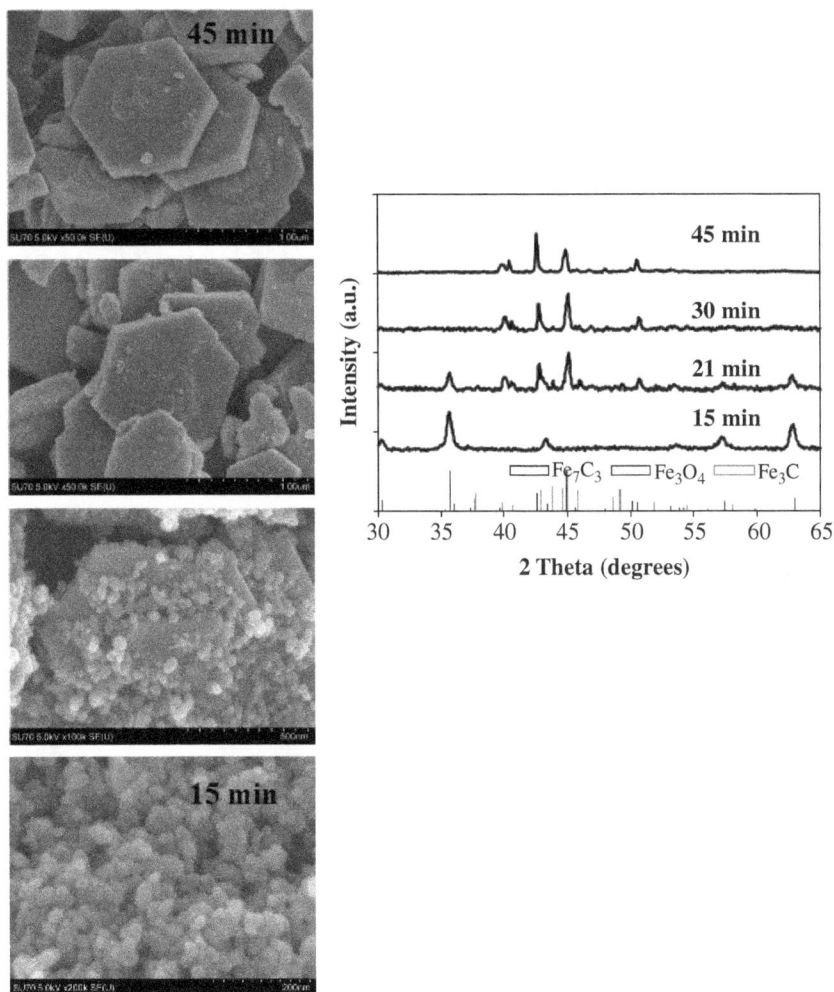

Figure 2.8: Example of changes in intensity peak ratios induced by preferential growth in one direction. A Fe carbide was processed for increasing times at $365°C$. XRD shows a mixture of phases in the initial processing stages. With increasing time, the Fe_3C phase becomes predominant. The peak at $42°$ becomes prominent because of the preferential growth of the base of the hexagonal prisms. Courtesy B. Williams, E. Carpenter.

2.1.7. *The X-Ray diffractometer*

The working principle of an X-ray diffractometer (XRD) will now be described. First off, let us revisit the Bragg condition (Eq. 2.14) and notice that incident and scattered beams form an angle 2θ, not θ, as shown in Figure 2.9. The reflection geometry used in the derivation of Bragg's formula is also used in many commercial XRDs, where the source and the detector swing on a circle, as shown in Figure 2.10. This arrangement of source and detector, known as the Bragg-Brentano configuration, helps with beam focusing. Beam focusing, which is not going to be discussed here, increases the intensity at the detector. For this reason the Bragg-Brentano configuration is very popular, which explains why angles are usually reported as 2θ in XRD plots, see for example Figure 2.7. Figure 2.9 also shows one

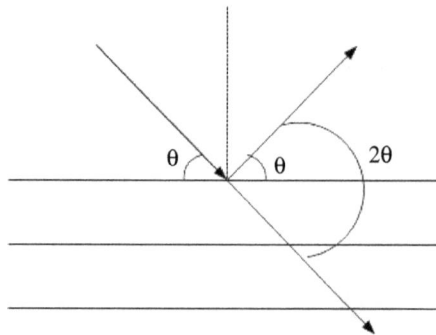

Figure 2.9:   Bragg's law, revisited.

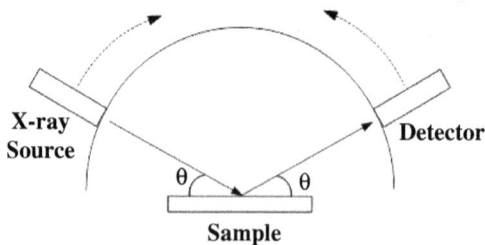

Figure 2.10:   Bragg-Brentano XRD scattering geometry.

limitation of the Bragg-Brentano configuration. One will get reflections into the detector only when the planes of the solid phase have an orientation that allows them to act as mirrors. For example, one would not get any reflection from a set of planes oriented at 45° from the surface normal, as shown in Figure 2.11. To get reflection from those planes, one would have to rotate the sample. This limitation explains why XRD is not used for single crystals, at least in the Bragg-Bentano configuration. XRD, however, is a very powerful tool for powder analysis. In powders, one deals with multiple crystallites oriented at random, as shown in Figure 2.12. The spectrum of powders yields diffraction from ALL crystalline planes, but

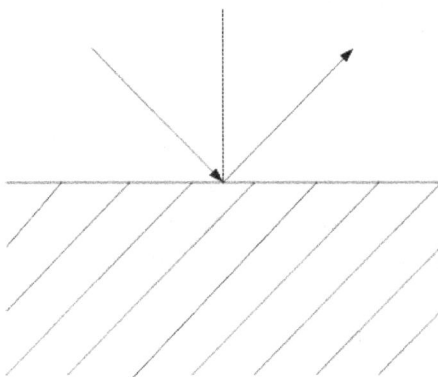

Figure 2.11:   The Bragg-Brentano configuration will not work for a single crystal with this orientation!

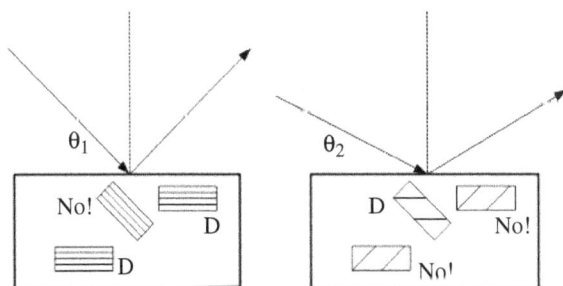

Figure 2.12:   X-ray diffraction from powder samples. At angle θ_1, only the planes of the particles indicated by D will contribute to diffraction. At angle θ_2, the situation has reversed.

the reflections come from different particles; see Figure 2.12. However, if the sample does not contain a sufficient number of particles, weird things can happen; the most annoying (and dangerous) of which being to have certain peaks be more intense than others simply because a larger number of particles in the sample are by chance in the "right" orientation. Or, one could have two different phases, or two different materials. If one phase were more abundant than the other simply because of a statistical fluctuation in the sample, one could erroneously conclude that one phase were more abundant than the other.

This Section will now be concluded by some technical remarks on the X-ray source, the detector, and the collimation system. The X-ray source is schematically shown in Figure 2.13. It consists of a chamber kept under vacuum, which hosts a filament heated by a DC current. The filament is at a higher potential than the anode (target). As a consequence, electrons (black arrows in Figure 2.13) are accelerated towards the target, which emits X-Rays (white arrows). X-rays are emitted in all directions, and only a small fraction will go in the desired direction. To prevent absorption of the X-rays a Beryllium window is used and placed at the desired angle of extraction. Be has a low atomic number and therefore is not a strong X-ray absorber. Since it is a metal, it is mechanically strong and it can be

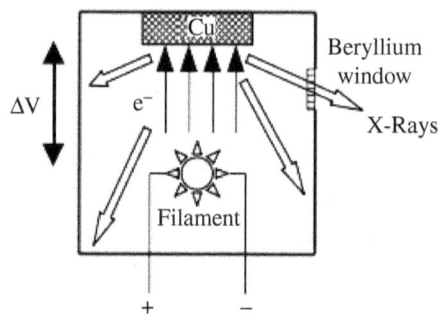

Figure 2.13: Schematic representation of an X-Ray source. Note how X-rays are emitted in all directions.Only a tiny fraction will escape through the Be window. Not shown is the massive water cooling system necessary to prevent melting of the Cu target and other parts of the assembly.

machined to a very low thickness without collapsing under the pressure differential. It is, however, a hazardous material and it should be handled with proper precautions. It should also be noticed that X-ray production by electron bombardment is a grossly inefficient process, which results in massive heat production. Hence, large currents must be directed at the target in order to produce acceptable beam intensities. Typical source powers are between 1 and 10 kW and they induce a considerable heating of the target. For good measure, the target is under vacuum; therefore air convection is not there to help with cooling. To prevent melting of the target, X-ray sources must be cooled with copious amounts of water. It is therefore not surprising that one of the most conspicuous features of a commercial XRD is a set of large ($\sim$2 cm diameter) water cooling pipes.

It must now be observed that the X-Ray beam exiting the source is still highly divergent. Angular divergence is undesirable, since Bragg's formula has a strong dependence on the angle of incidence. To reduce divergence, the beam is passed through a series of slits, as shown in Figure 2.14a), or through a parallel-plate collimator, shown schematically in Figure 2.14b). Collimation further reduces the beam intensity, but it is a necessary evil. To detect the scattered

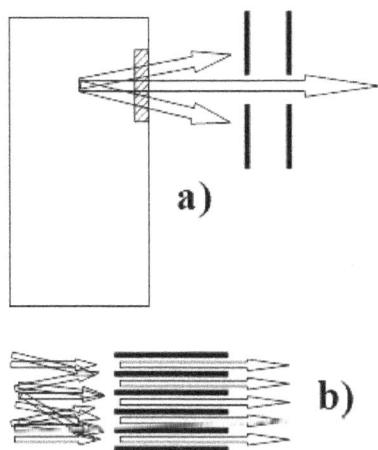

Figure 2.14: Reduction of beam divergence using (a) Slits. (b) Parallel plate collimator.

X-rays, the most popular schemes are charged coupled devices (CCD) or semiconductor strips. CCDs measure the charge accumulated on each pixel, much like digital cameras. They are becoming less popular because of cost, but also because of high noise. A more popular detector is based on semiconductor strips. These are diodes operated at high voltage ($\sim$100 V) and reverse polarity. Without irradiation, only a small current flows through these diodes. When ionizing radiation interacts with the detector, it produces electrons and holes, which migrate to the cathode and anode respectively, and generate a current pulse. A discriminator separates noise from "true" pulses and increases the signal-to-noise ratio.

Let us now put these things together:

- The incident beam has a small, but nonzero, divergence.
- The scattered beam has also some divergence (collimators and slits are used also for the scattered beam, but some divergence is unavoidable).
- The detector has a finite area; therefore it introduces some indetermination on the angle.
- The goniometer which controls the source and detector is not infinitely precise.
- The incident X-rays are not totally monochromatic.

All these factors introduce an angular uncertainty. This is why peaks from large particles and/or bulk samples are not infinitely narrow, but have a certain (minimum) width, typically of around 0.2 degrees. This intrinsic line width is sometimes neglected in the analysis of XRD data, leading to errors in the interpretation of the spectra, especially regarding particle size.

2.2. Small angle scattering techniques

These techniques (mostly X-Ray, but also neutron) are very useful to determine the long-range structural order and spatial arrangement of nanoparticles in a composite. The key observation is to recognize that in Bragg's law, Eq. (2.14), the angle becomes smaller when the interplanar spacing, d, increases. Therefore, we should be able to observe periodicities on the order of tens of nm, if we could measure scattering at small angles. We call these techniques Small Angle X-ray Scattering (SAXS) and SANS for neutron scattering.

Figure 2.15 is key to the theoretical treatment. We have a scatterer (which can be anything: a particle, or an electron, or an atom) located at point O and another scatterer located at point P. A plane wave interacts with the scatterers and we look at the wave scattered at an angle θ.

Scattering is an interference problem, so we need the phase difference between the waves scattered at O and respectively P. This phase difference is $\Delta\phi = \frac{2\pi\delta}{\lambda}$, where δ is the geometrical path difference. To make life easy, we assume that scatterer O is at the origin of the coordinate system. P will be at position $\mathbf{r}$, $\bar{\mathbf{QP}} = \mathbf{S_0} \cdot \mathbf{r}$ and $\bar{\mathbf{OR}} = \mathbf{S} \cdot \mathbf{r}$, and therefore $\Delta\phi = \frac{2\pi}{\lambda}(\mathbf{S_0} \cdot \mathbf{r} - \mathbf{S} \cdot \mathbf{r}) = -2\pi\mathbf{S} \cdot \mathbf{r}$, where $\mathbf{s} = \frac{\mathbf{S} - \mathbf{S_0}}{\lambda}$. $\mathbf{s}$ is called the scattering vector. The scattering vector is a particularly relevant quantity, since it defines the phase difference and therefore the scattering intensity. In many theoretical treatments, instead of $\mathbf{s}$ the vector $\mathbf{q}$ is used, which is defined as $\mathbf{q} = 2\pi\mathbf{s} = \mathbf{k} - \mathbf{k_0}$.

Now, the incident wave is $A_0 e^{i(\omega t - \mathbf{k} \cdot \mathbf{r})}$, $k = \frac{2\pi}{\lambda}$. The wave scattered at O is: $A_1 = A_0 b e^{i(\omega t - \mathbf{k} \cdot \mathbf{r})}$, where b is the scattering length (i.e., the scattering cross-section). The wave scattered at P is

$$A_2 = A_1 e^{i\Delta\phi} = A_0 b e^{i(\omega t - \mathbf{k} \cdot \mathbf{r})} e^{-i2\pi\mathbf{s} \cdot \mathbf{r}}. \tag{2.19}$$

The wave at the detector will be the sum of the two scattered waves: $A = A_1 + A_2$. Since the detector measures the energy of the

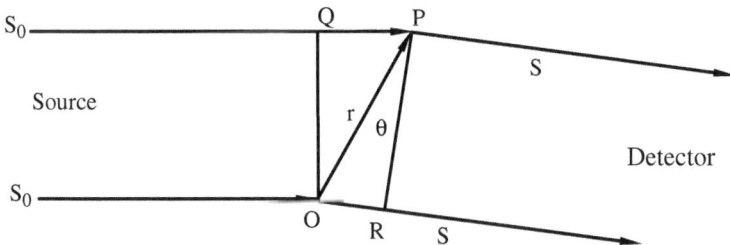

Figure 2.15: Scattering geometry.

waves, the scattered intensity is:

$$I = AA^* = A_0^2 b^2 (1 + e^{-i2\pi \mathbf{s} \cdot \mathbf{r}})(1 + e^{i2\pi \mathbf{s} \cdot \mathbf{r}}). \qquad (2.20)$$

We notice that the parts containing time and position vanish in Eq. (2.20). This makes our life easier, since we can neglect these dependencies throughout the discussion. That is, Eq. (2.19) can be written as:

$$A(\mathbf{s}) = A_0 b (1 + e^{-i2\pi \mathbf{s} \cdot \mathbf{r}}). \qquad (2.21)$$

If we have N scatterers and not only two:

$$A(\mathbf{s}) = A_0 b \sum_{j=1}^{N} e^{-i2\pi \mathbf{s} \cdot \mathbf{r}_j},$$

or, when the material can be treated as a continuum:

$$A(\mathbf{s}) = A_0 b \int \rho(\mathbf{r}) e^{-i2\pi \mathbf{s} \cdot \mathbf{r}} d\mathbf{r} = A_0 b \int \rho(\mathbf{r}) e^{-i\mathbf{q} \cdot \mathbf{r}} d\mathbf{r}, \qquad (2.22)$$

where $\rho(\mathbf{r})$ is the density of scatterers.

Equation (2.22) is the master equation for SAXS. If the scatterers have a simple shape and are monodisperse, Eq. (2.22) can be calculated analytically. For example, if the scatterers are monodisperse spheres of radius R: $\rho(\mathbf{r}) = \{ \begin{smallmatrix} \rho_0 & r \geqslant R \\ 0 & r > R \end{smallmatrix}$, then:

$$A(q) = \rho_0 v \frac{3(\sin(qR) - qR\cos(qR))}{(qR)^3}, \qquad (2.23)$$

v being the volume of the particle. The scattered intensity is plotted in Figure 2.16. SANS was used in the distant past to determine the structure of viruses, which were approximated by spheres or spherical shells. See, for example, [3] and [4].

We note in Eq. (2.23) that the oscillations have a $1/q^3$ damping. Damping at large q is not limited to scattering from spheres. It is a characteristic of SAXS and it provides a considerable amount of information on the morphology of the system. Analytical calculations show that for large q, $I(q) \propto q^{-\alpha}$, with $\alpha = 4$ for spheres, $\alpha = 2$ for disks and $\alpha = 1$ for rods. If the scatterer is disordered, such

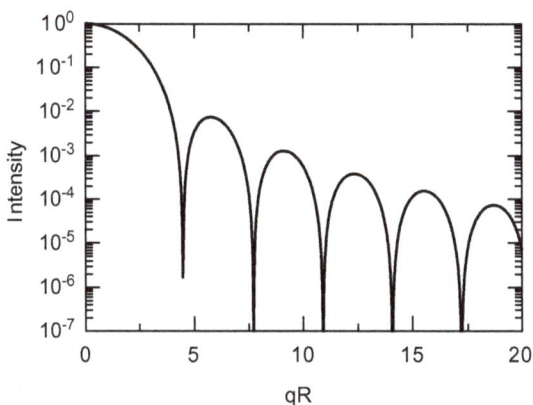

Figure 2.16: Scattering intensity for a solid sphere of radius R.

(a)   (b)

Figure 2.17: (a) Schematic representation of the Gaussian Approximation for a polymer. Beads of volume V and density r are arranged randomly and separated by a distance l. (b) A cauliflower, showing how small particles can coalesce into larger aggregates with different length scales.

as a polymer coil, the typical approximation is to assume the coil to be made of spherical beads, placed a distance l from each other (Gaussian approximation), see Figure 2.17. This approximation also yields a q^{-2} dependence, which can make the analysis confusing (are they disks or polymer coils?).

We also note that the small q part of the scattering profile yields information relative to the aggregation of particles on a large scale. The analogy is that of a cauliflower, where small aggregates coalesce into larger ones. For small q, another important expression (Guinier approximation) can be obtained:

$$I(q) = \rho_0^2 v^2 e^{(-1/3q^2 R_g^2)}, \qquad (2.24)$$

where R_g is called the gyration radius and it represents the size of aggregates into which nanoparticles coalesce. In a scattering experiment, the intensity will follow different trends as a function of q. Each region will provide information over a certain length scale of the system. For example, in Figure 2.18 at large q we notice a $1/q$ dependence, indicating that the smallest scatterers have a rod-like shape. We then observe a $1/q^2$ dependence, which can be reconciled with nanoparticles coalescing into a polymer. The polymer coils in turn coalesce into spherical aggregates of radius R_g (this is the small q region). The crossovers between these trends are related to the dimensions of the aggregates.

A good and pioneering example of the use of small angle scattering for the structural analysis of nanocomposite materials is reported in [5]. The nanocomposites were silica aerogels. Aerogels consist of

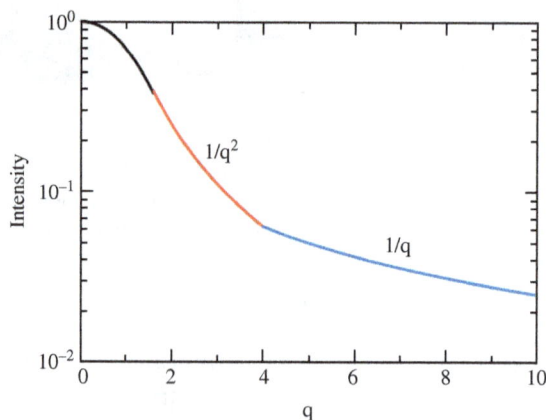

Figure 2.18: Example of rod-shaped particles (blue trace) coalescing into strand-like, polymeric aggregates (red trace) which in turn coalesce into spherical aggregates that can be reproduced by the Guinier approximation (black trace).

dense, spherical SiO_2 nanoparticles, which coalesce into secondary aggregates. In this work, the scattering of colloidal suspensions of spherical SiO_2 was compared with that of silica aerogels. The results are presented in Figure 2.19. Note that intensity and scattering vectors are both reported in logarithmic scale, which facilitates comparison with theory. At large q (k in the notation of the authors), all curves have a -4 slope, indicating that the smallest aggregates are spherical. For aerogels, a change to a -2 slope is observed at small q. This exponent indicates formation of chainlike aggregates reminiscent of those of branched polymers, as shown in Figure 2.19(b). The crossover is around 1.1 nm, which is the radius of the primary nanoparticles. Over the years, a number of theoretical models have been developed for the small angle scattering techniques, a very popular one being the Unified Beaucage Model [6]. These models (for which calculators and fitting algorithms are freely available on the web), together with the introduction of small footprint instrumentation, make SAXS an appealing technique for the characterization of nanocomposites.

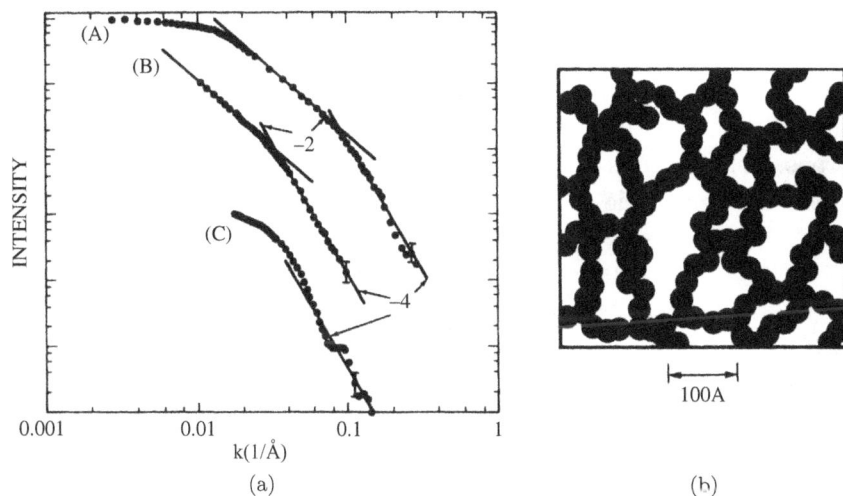

Figure 2.19: (a) Small Angle X-Ray scattering from aerogels of different densities (A, B) and colloidal suspensions of silica nanoparticles (C). (b) Proposed aerogel structure. Adapted with permission from [5]. © American Physical Society.

2.3. Electron microscopies

To understand why electron microscopy is so powerful, we start by remembering that the resolution δ of a microscope depends on the wavelength of the radiation via the Abbe relationship:

$$\delta = \frac{0.61\lambda}{n\sin\theta}, \qquad (2.25)$$

where λ is the wavelength of the incident light, n the index of refraction and θ the collection angle (actually, half the collection angle). For visible light the Abbe limit (using glass lenses) is on the order of 300 nm. For electrons, Eq. 2.25 still holds (more or less), but now λ depends on electron energy via the De Broglie relationship:

$$\lambda = \frac{h}{\sqrt{2Em}} \quad \text{or} \quad \lambda = \sqrt{\frac{1.5}{V}}\text{nm}, \qquad (2.26)$$

where V is the accelerating voltage and m is the mass of the electron. For energies on the order of 100 keV, Eq. (2.26) yields λ in the pm range, that is, well below the interatomic spacing in solids; see also Table 2.1.

2.3.1. *Electron lenses*

Microscopy (optical or electronic) requires lenses. Lenses are all about changing the direction of particles, be they photons or electrons. For electrons, one has two ways of changing their trajectory: electric or magnetic fields. Deflection by electric fields makes use of

Table 2.1: Electron wavelengths as a function of the accelerating voltage. Relativistic corrections have been omitted. Note how the changes in wavelength decrease with accelerating voltage.

Voltage (kV)	λ(nm)
100	3.87×10^{-3}
200	2.73×10^{-3}
300	2.23×10^{-3}

a parallel-plate condenser, as shown in Figure 2.20. To calculate the deflection, it is assumed that an electron passes through two plates of length L separated by a distance d. The initial velocity of the electron is v, and a voltage V is applied across the plates, which generates an electric field $\mathbf{E}$. Using elementary kinematics, and remembering that the transit time through the capacitor is $t = \frac{L}{v}$, we obtain: $a_x = \frac{e|E|}{m}$, and $v_x = \frac{e|E|}{m} \cdot \frac{L}{v}$.

For a magnetic field, the force on the electron is the Lorentz Force: $\mathbf{F} = -e\mathbf{v} \times \mathbf{B}$. For small deflections, the force can be assumed parallel to the x-axis (see also Figure 2.21), and therefore: $|a_x| = \frac{evB}{m}$, and $v_x = |a_x|t = \frac{evB}{m} \cdot \frac{L}{v} = \frac{eBL}{m}$. The angle of deflection is calculated (for small deflections and neglecting relativistic corrections) as:

$$\theta = \begin{cases} \dfrac{eEL}{mv^2} & \text{Electrostatic field,} \\[2mm] \dfrac{eBL}{mv} & \text{Magnetic field.} \end{cases} \tag{2.27}$$

The ratio between the deflection θ_E in and electrostatic field and the deflection θ_B in a magnetic field is then:

$$\frac{\theta_E}{\theta_M} = \frac{eEL}{mv^2} \times \frac{mv}{eBL} = \frac{E}{Bv}. \tag{2.28}$$

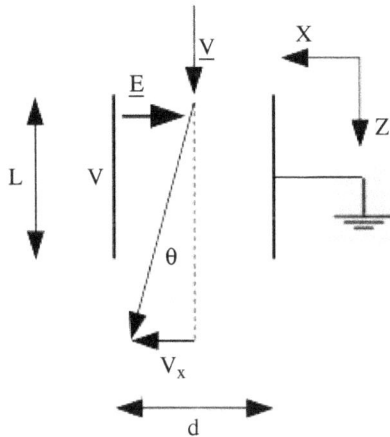

Figure 2.20: Deflection of an electron by an elecrostatic field.

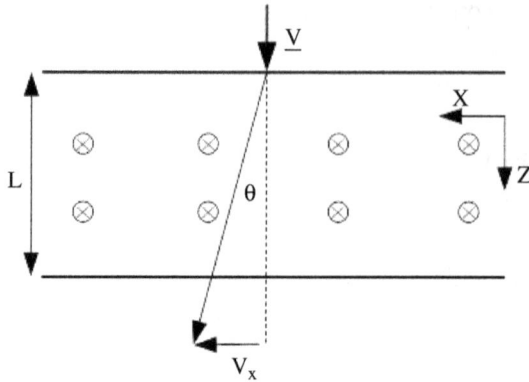

Figure 2.21:   Deflection of an electron in a magnetic field.

Let us now assume that the electrostatic and magnetic deflections are equal. The right-hand side of Eq. (2.28) becomes unity, and therefore:

$$E = Bv. \tag{2.29}$$

Equation (2.29) tells us which value the electric field E must have in order to equal the deflection in a magnetic field B. The problem is that electrons travel quite fast inside electron microscopes. At 100 keV, the velocity of an electron is $v \sim 10^8$ m/s, therefore E must be $\sim 10^8$ V/m to attain the same deflection that we get in a $B = 1$ T field. Such electric high fields are not practical; this explains why magnetic lenses are preferred to electrostatic lenses.

2.3.2.   *Trajectories inside an electron lens and image formation*

Magnetic lenses in an electron microscope are typically obtained by winding a coil inside a donut-shaped pole piece, as shown in Figure 2.22. For this configuration, the field on the vertical (z) axis is approximated reasonably well by a bell-shaped curve:

$$B_z = \frac{B_0}{1 + \frac{z^2}{a^2}}. \tag{2.30}$$

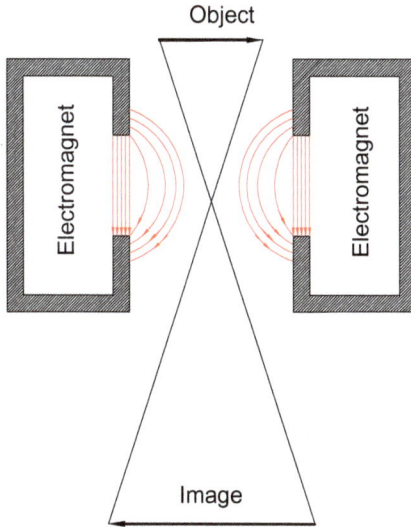

Figure 2.22: Schematic representation of a pole piece, of the field inside it (red lines) and of image formation.

One can also demonstrate that the radial field is given by:

$$B_r = -\frac{r}{2}\frac{\partial B_z}{\partial z}.$$

(2.31)

Let us now look at what happens to an electron moving inside such a field. Working in cylindrical coordinates, we obtain:

$$m\ddot{r} = F_r + mr\dot{\phi}^2$$
$$\frac{d}{dt}(mr^2\dot{\phi}) - rF_\phi$$
$$m\ddot{z} = rF_z$$

(2.32)

where F_r, F_ϕ are the radial and circular components of the force acting on the electron, respectively. This is force is none other than the Lorentz force: $\mathbf{F} = -e\mathbf{v} \times \mathbf{B}$, with $\mathbf{v}$ in cylindrical coordinates being $\mathbf{v} = (\dot{r}, r\dot{\phi}, \dot{z})$. The Lorentz force can be plugged into Eq. (2.32) to get:

$$m\ddot{r} = -eB_z r\dot{\phi} + mr\dot{\phi}^2$$

$$\frac{d}{dt}(mr^2\dot{\phi}) = eB_x r\dot{r} + e\frac{r^2}{2}\dot{z}\frac{\partial B_z}{\partial z} = \frac{d}{dt}\left(\frac{e}{2}r^2 B_z\right) \qquad (2.33)$$

$$m\ddot{z} = eB_r r F_z$$

Key to this system of equations is the second equation. Its integration yields:

$$mr^2\dot{\phi} = \frac{e}{2}r^2 B_z + C, \qquad (2.34)$$

where C is an integration constant. This integration constant is pesky, because it hides the physics of the problem. To get rid of it, let us consider a ray that goes through the center of the lens. For that ray, $r = 0$, and Eq. (2.34) yields $C = 0$. With C removed from the picture, Eq. (2.34) becomes: $\dot{\phi} = \frac{eB_z}{2m}$, which means that the electron will move in a spiral along the lens axis with an angular velocity (Larmor frequency) $\omega_L = \dot{\phi}$.

Let us now introduce the "true" form of the magnetic field, Eqs. (2.30) and (2.31), into the equations of motion, Eq. (2.33). The mathematical treatment is rather tedious and only its conclusions for paraxial rays (i.e., r small, that is, rays close to the lens axis) will be discussed. We first have to introduce new coordinates and a parameter:

$$y = \frac{r}{a}$$

$$\cot(\chi) = \frac{z}{a} \qquad (2.35)$$

$$k = \frac{eB_0^2 a^2}{8m_0 \tilde{U}}$$

In Eq. (2.35), y is the radial part; χ is related to the position of the electron within the lens. When the electron is far away from the lens, $z = -\infty$ and $\chi = \pi$. When the electron is far away from the lens and has crossed it, $z = +\infty$ and $\chi = 0$. The parameter k is related to the field strength B_0. $\tilde{U} = U(1 + \frac{E}{2E_0})$, where U is the accelerating voltage, $E = eU$ the kinetic energy and $E_0 = m_o c^2$ is the energy of the electron at rest. And yes, the relativistic correction is being used in this case. Using these variables and parameters, an equation for y (i.e., the radial coordinate) is obtained:

$$y(\chi) = \frac{\sin(\omega\chi)\sin(\chi_0)}{\sin(\omega\chi_0)\sin\chi} y_0 + \frac{D}{\sin\chi} \left[\cos(\omega\chi) - \frac{\cos(\omega\chi_0)}{\sin(\omega\chi_0)}\sin(\omega\chi) \right].$$

$$(2.36)$$

In Eq. (2.36), $\omega = 1 + k^2$, and y_0 expresses the distance between the lens axis and the initial position of the electron. If the electron were to enter on the axis, $y_0 = 0$, but in general $y_0 \neq 0$. χ_0 is related to the initial distance z between the electron and the lens. χ_0 is not strictly π because the distance z is finite. In addition, in electron microscopes there are multiple lenses; hence χ_0 will depend on the lens configuration specific of a microscope. There is another initial condition, and that is the direction of the incident electron: y_0 and χ_0 tell us where the electron enters the lens, but not at which angle. The incident direction is given by the constant D in Eq. (2.36). To better understand the role of D let us take a step back and consider an optical lens, an example of which is reported in Figure 2.23. In this example, the image forms on the plane where rays emitted by the same point of the object, but in different directions, converge. Thus, the conjugate point (that is, the image point) of the original

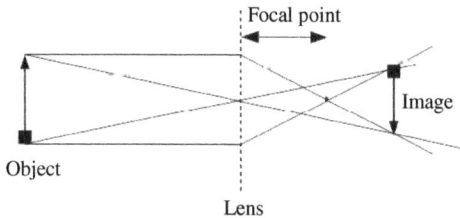

Figure 2.23: Formation of an image in optical systems. Rays emitted by the object in different directions converge to the same plane.

point of an object is independent on the direction of the rays emitted by the object. That is, the image plane is independent on D.

For that to be true, the square bracket in Eq. (2.36) must be zero:

$$\cos(\omega\chi) - \frac{\cos(\omega\chi_0)}{\sin(\omega\chi_0)}\sin(\omega\chi) = 0. \tag{2.37}$$

A bit of math shows that the axial position χ_1 of the conjugate point is:

$$\chi_1 = \chi_0 - \frac{n\pi}{\omega}, \quad \text{n integer}, \tag{2.38}$$

and that the radial position y_1 of the conjugate point is:

$$y_1 = \frac{\sin(\omega\chi)\sin(\chi_0)}{\sin(\omega\chi_0)\sin\chi}y_0 = My_0 \tag{2.39}$$

Equations (2.38) and (2.39) tell us where the image will be formed. The image will be magnified by a factor M which is derived in Eq. (2.39). Also, we note that Eq. (2.38) has multiple solutions, i.e., the image can form in more than one place. This calamity, however, seldom occurs in electron microscopes! Before we close, it must be emphasized that magnetic lenses have many aberrations. This is why electron microscopes must be properly aligned (i.e., aberrations minimized) before use. A more extensive analysis of image formation in electron microscopes can be found in Ref. [7].

2.3.3. *The scanning electron microscope (SEM)*

In an SEM, an electron beam with an energy between about 5 and 15 keV is focused very tightly (down to 1-2 nm) onto a specimen. SEM analyzes electrons and X-rays scattered back from the specimen into the vacuum and provides, for the most part, information on the surface topology and the chemical composition of the specimen. Images are obtained by scanning the beam over the sample, which can have any thickness. Incident electrons interact with the specimen in a pear-shaped region, as shown in Figure 2.24. The interaction with the specimen leads to three different signals:

a) Secondary electrons (SE). These are electrons that are removed from the sample. They have a low energy (<50 eV) and therefore they travel only for very short distances within a solid (a few nm). Thus, only SE produced near a surface will be able to reach a detector.

b) Backscattered electrons. These are incident electrons that lost some energy to the sample before being scattered back into vacuum. The energy of backscattered electrons is much higher than that of the secondary electrons, and therefore they can travel distances of a few μm inside the sample. Because of their comparatively large interaction region, backscattered electrons provide information with lower spatial resolution than secondary electrons.

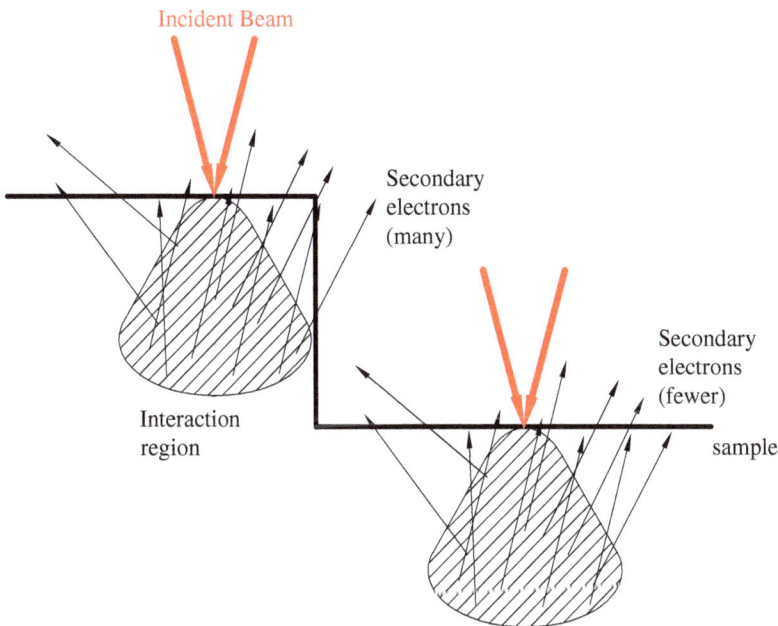

Figure 2.24: Interaction region of electrons with a specimen in SEM and origin of contrast.

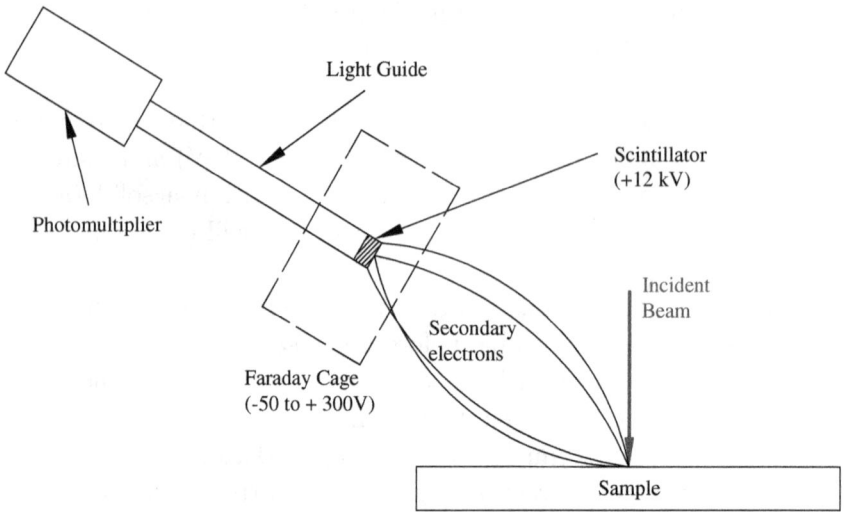

Figure 2.25:   Detection of SE in a SEM.

c) X-Rays. These are emitted when the incident electrons remove an electron from the inner shell of a substrate atom. X-Rays provide information on the chemical composition of the specimen.

SEM imaging is all about the secondary electrons. They are collected from the sample by placing a Faraday cage around a scintillator. The cage is held at a positive potential to attract the electrons and direct them onto a scintillator. Light emitted by the scintillator is then collected by a photomultiplier. This Everhardt-Thornley detector is shown in Figure 2.25. Contrast and topography derive from the different interaction volumes of different regions, as shown in Figure 2.24, and the resolution is given by the size of the incident beam.

2.3.4. *The transmission electron microscope (TEM)*

The working principle of a TEM is shown in Figure 2.26. A beam is produced by an electron gun and collimated onto the sample via a set of condenser lenses. The parallel (not convergent!) beam interacts with the specimen and is then magnified by a series of projector

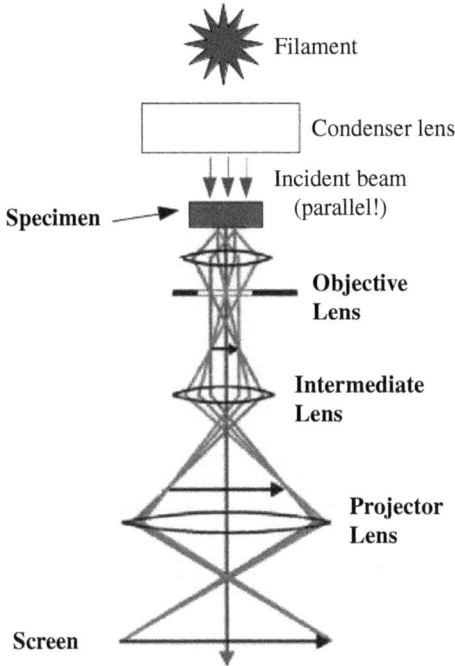

Figure 2.26:   TEM in imaging mode.

lenses. These lenses allow magnifications of several hundred thousand times.

Key to imaging is the interaction between electrons and the specimen, which is schematically shown in Figure 2.27. Some electrons (secondary electrons) are reflected back, and we have seen their utility when discussing SEM. Other electrons get through the sample but are scattered inelastically in all directions. Other electrons are diffracted and some are transmitted. There are other interactions, but we will limit the discussion to diffracted and transmitted electrons.

The transmitted electrons are what one usually works with in TEM. This imaging mode, called bright field, is illustrated in Figure 2.28. Fewer electrons will be transmitted through thicker regions of the sample, which will appear darker on the screen or camera. That is, in TEM we see "shadows" of the specimen.

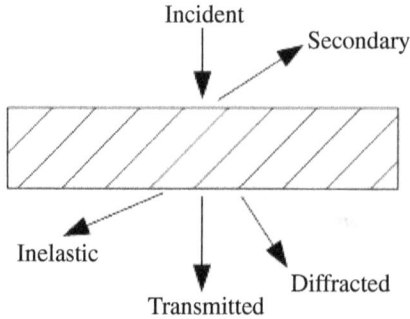

Figure 2.27:   Interaction of an electron beam with a specimen in TEM.

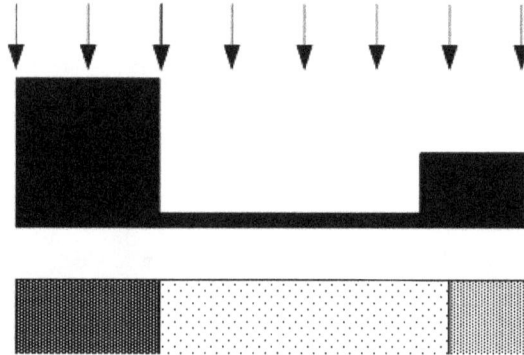

Figure 2.28:   Contrast in TEM. Thicker regions (left) transmit fewer electrons and appear darker than thinner regions on the screen/camera.

The darkness of a feature depends on its thickness but also on the atomic number of the constituent atoms.

Another powerful way of operating the TEM is in diffraction mode. Diffracted electrons give rise to patterns on the screen in the same fashion as X-Ray diffraction patterns on a photographic plate. We note that the electrons will be diffracted at some angle from the normal. For these electrons to reach the screen, the magnification of the microscope must be reduced, as shown in Figure 2.29. Figure 2.30 shows this mode of TEM operation, which is called Selected Area

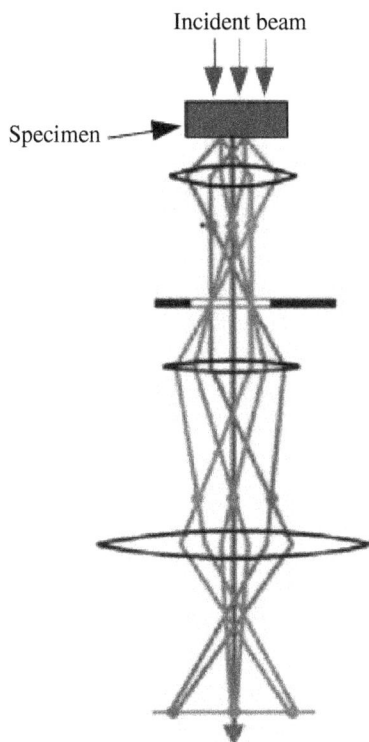

Figure 2.29: TEM operated in diffraction mode.

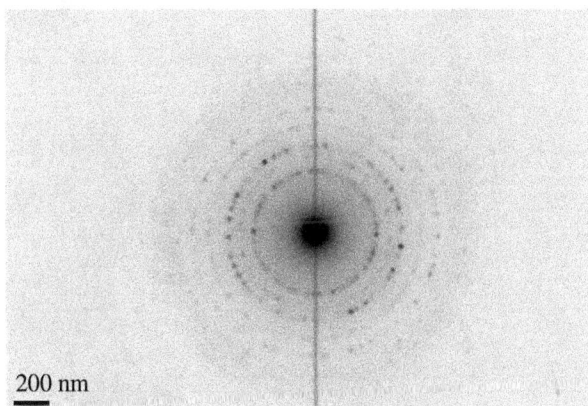

200 nm

Figure 2.30: Example of Selected Area Diffraction (SAED) from BaSrTiO$_3$ nanoparticles. Adapted with permission from JEOL.

Electron Diffraction (SAED). We note diffraction rings. The interplanar spacing can be calculated from the diameter of the rings,
similar to what is done with the scattering angle in XRD. We also
note a bright region in the center of the SAED pattern. This region
is actually a low magnification image of the specimen. It appears
bright because most electrons are transmitted through the sample.
Diffracted electrons represent only a small percentage of the total
beam intensity. To prevent damage to the camera, the central part
of the beam is typically masked off by a beam stop.

Another mode of operation of a TEM is high resolution TEM
(HREM). Today's TEMs allow routinely to image lattice fringes
when operated in HREM mode, and an example is shown in the
upper panel of Figure 2.31. Before we proceed, let us note that lattice
fringes are not an image of the lattice planes, but rather the diffraction image of the lattice planes. This is perhaps a small technical
detail but one must bear this in mind. The periodicity of the fringes
can be measured directly on the image, and from it, the interplanar
spacing can be calculated. Alternatively, one can take the Fourier
transform of the part of the image containing the lattice fringes and
calculate the interplanar spacing from the power spectrum. There

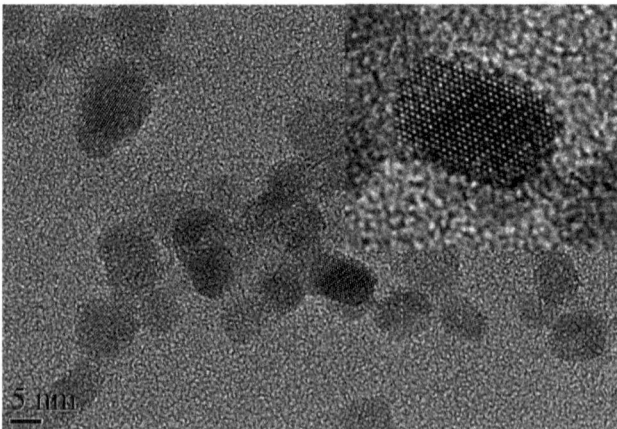

Figure 2.31: Example of HREM from BaSrTiO$_3$ nanoparticles. Adapted with
permission from JEOL.

is a caveat, however. Only a limited number of lattice fringes can be typically observed in HREM. Those fringes will come from those planes whose orientation is "right", much like in the single-crystal XRD example of Figures 2.12 and 2.11. Since it is very difficult to calculate lattice constants and crystal symmetry from a couple of lattice spacings, we emphasize that HREM gives only partial information on the crystalline habit of a particle. SAED, instead, provides the lattice constant and symmetry, and is therefore preferable to HREM for crystallographic purposes. SAED is also more rapid and can be carried out on much cheaper microscopes than HREMs and is therefore more time- and cost-effective. HREM, instead, is invaluable to evaluate interfaces, defects and surfaces, as shown in Figure 2.32.

2.4. Scanning tunneling microscopy (STM)

STM is based on quite a simple principle: the quantum tunnel effect. When a metal tip is brought in proximity of a surface and a voltage is applied, electrons can tunnel into (or from) the tip. The tunneling current provides information on the morphology and the energetic levels of the surface. The device is schematically represented in Figure 2.35: an atomically sharp tip is attached to three piezoelectric

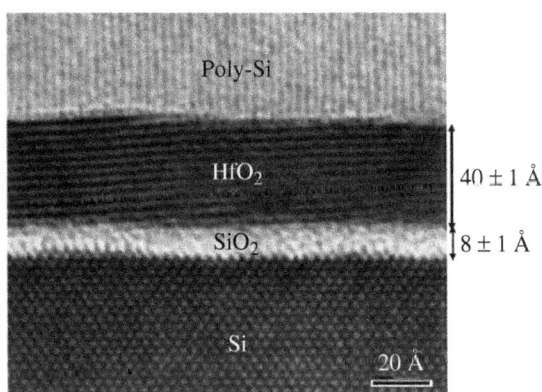

Figure 2.32: Example of interface analysis via HREM. Adapted with permission from JEOL.

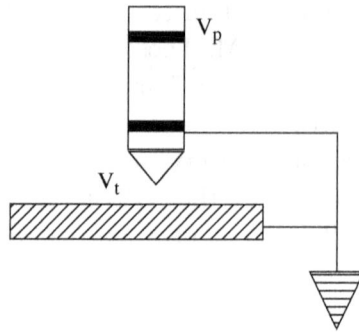

Figure 2.33: Schematic representation of an STM. An atomically sharp tip is connected to three piezoelectric transducers (only the vertical one being shown here) and kept at a voltage V_t with respect to the sample (grounded). The shape of the piezoelectric transducer is changed by applying a voltage V_p across it.

tubes (one for each axis). A voltage V is applied between tip and sample, the latter being usually kept at ground. The piezoelectric components are key to the operation of the STM. A voltage is applied across them, as shown in Figure 2.33, and this voltage induces a change in length:

$$\Delta L = CV, \tag{2.40}$$

where C is a constant (piezoelectric coefficient) which is characteristic of the material. For materials such as $PbTiO_3$, C is on the order of $\text{Å}/V$, which explains why the tip can be controlled with such accuracy. The piezoelectric transducers, however, are not the real reason for the high resolution of an STM. The core of this technique is the tunneling of electrons from the sample to the tip. The theoretical framework was developed in pioneering work by Tersoff *et al.* [8]. The authors considered the case of a particle of mass m and energy $E < U$, where U is the height of the energy barrier between the energetic levels of the tip and the sample, as shown in Figure 2.34.

For such a system, the electron wavefunction has the form:

$$\psi(z) = \psi(0)e^{(-kz)}, \text{ where } k = \sqrt{\left(\frac{2m(U-E)}{\hbar}\right)}. \tag{2.41}$$

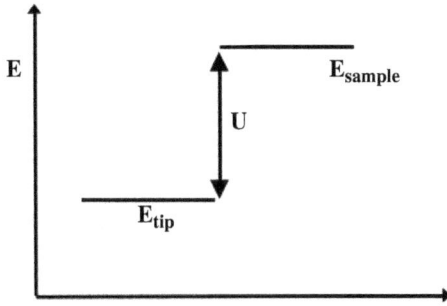

Figure 2.34:   Schematic representation of the energy diagram used for theoretical treatment of STM.

The tunneling probability is:

$$|\psi(z)|^2 \propto e^{(-2kz)} \qquad (2.42)$$

which decreases very rapidly with distance. This strong dependence on z explains why STM is so sensitive to the surface morphology. To make the treatment more quantitative, we observe that the energy barrier is on the order of the work function ϕ of a metal, which expresses the energy mecessary to move an electron from the Fermi level to vacuum. For most metals and semiconductors, $\varphi \sim 5$ eV. Therefore, for small bias E, $k \sim 1\text{Å}^{-1}$, which quantifies and explains the rapid decay of the tunneling current with the distance between the tip and the surface. The lateral resolution of an STM (which is as important as the vertical resolution) can be calculated starting from the geometry of Figure 2.35. We assume the tip to be monoatomic and the tip atom to have an s-type wave electron density:

$$|\psi(r)|^2 \propto \frac{e^{(-2kr)}}{r^2}, \text{ where } r = \sqrt{(x^2 + z^2)} \qquad (2.43)$$

For $z \gg x$:

$$|\psi(x,z)|^2 \propto \frac{e^{(-2kz)}}{z^2}e^{(\frac{-kx^2}{z})}, \qquad (2.44)$$

which yields the now-familiar exponential decay in z multiplied by a Gaussian function. The lateral resolution of the STM is the full width at half maximum (*fwhm*) of the Gaussian: $\Delta x = \sqrt{(\frac{2z}{k})}$. We

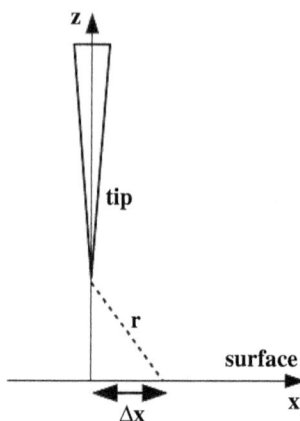

Figure 2.35:　Lateral resolution of an STM.

remember that $k \sim 1\mathring{A}^{-1}$; therefore the lateral resolution is on the order of $\mathring{A}$.

Because of its outstanding spatial resolution and ability of measuring the surface corrugation, STM was an incredibly popular technique between about 1985 and 2000. Interest then waned for a series of reasons, the most relevant of which being the failure to adapt STM to "real-world" surfaces (that is, contaminated and larger than a few millimeters).

A relevant application of STM, however, remains: scanning tunneling spectroscopy (STS). STS allows to probe the electronic states of a surface and is extremely relevant for nanotechnology applications [9]. The set-up of an STS experiment is shown in Figure 2.36. A tip is positioned in proximity of a surface, and the tunneling current is measured while the the tip bias is increased. In this configuration, the transmission coefficient T (i.e., the tunneling probability) can be expressed as [10]:

$$T \propto \exp\left(\frac{-2z}{\hbar}\sqrt{\left(m\left(\frac{(\phi_{tip} + \phi_{sample})}{2} + eV - E\right)\right)}\right) \qquad (2.45)$$

where z is the tip-surface distance, m the mass of an electron, φ the work function of tip and sample, respectively, and E is the energy of

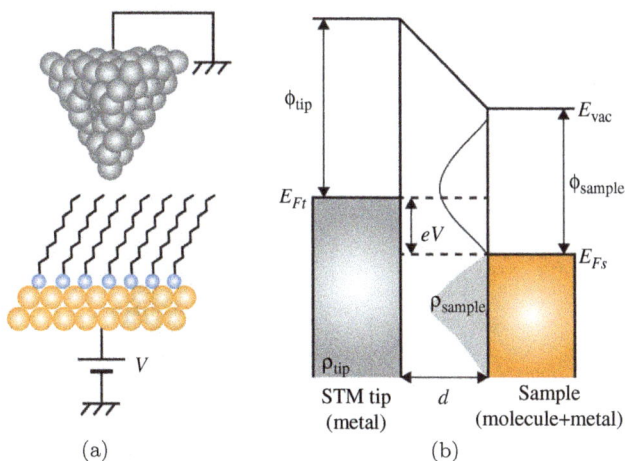

Figure 2.36: Schematic set-up of an STS experiment. Adapted with permission from [9]. © The Royal Society of Chemistry.

the electron. The tunneling current I is obtained by integrating the transmission coefficient while including the density of states ρ of tip and sample:

$$I \propto \int_0^{eV} \rho_{tip}(E - eV)\rho_{sample}(E)T(E, V, z)dE. \qquad (2.46)$$

A perhaps more useful relationship can be derived by assuming to work at a constant distance from the surface and low applied voltage:

$$\frac{dI}{dV} \propto \rho_{sample}(V)\rho_{tip}(0)T. \qquad (2.47)$$

That is, STS can probe the local density of states of a substrate. Equation (2.47), and the power of STS, can be best visualized for a system with discrete energetic levels, such as a molecule or a quantum dot [11], as shown in Figure 2.37. Peaks in the dI/dV curve will appear whenever the applied voltage raises the energy of the tip's electrons above one of the energetic levels of the substrate. We also observe that moving an electron from the tip to the nanoparticle induces charging of the nanoparticle. One thing to take into account

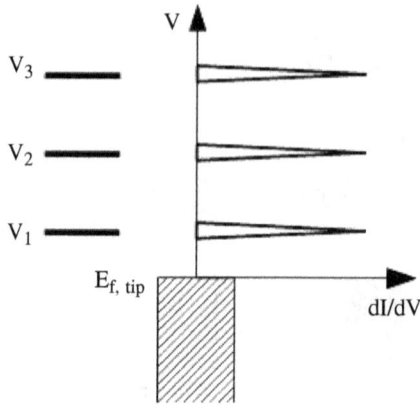

Figure 2.37: Use of STS on a system with discrete energy levels. The voltage V is increased and dI/dV exhibits a spike each time $E_{f,tip} + V$ matches one of the energy levels of the surface.

in STS is the electrostatic charging energy:

$$E_c = \frac{e^2}{2\epsilon r}. \tag{2.48}$$

Equation (2.48) represents the charging energy for a sphere in free space. Assuming $r \sim 1$ nm and $\varepsilon \sim 10$ (typical of semiconductors), $E_c \sim 100$ meV. The charging energy is an energetic barrier which manifests itself as plateaus in I-V curves and spikes in dI/dV curves and can complicate experimental analysis.

2.5. Atomic force microscopy (AFM)

In atomic force microscopy a laser is reflected off a tip holder into a position-sensitive photodetector, as shown in Figure 2.38. The information is provided by the deflection of the cantilever. AFM is characterized by a large number of imaging modes. However, one should remember that the tip interacts with the surface in first approximation via van der Waals forces. We will now describe the most popular imaging modes: contact and resonant modes.

Contact mode. This was the first imaging mode to be developed. In this mode, the tip is brought in close proximity of the surface where it interacts with the repulsive part of the potential. In contact mode,

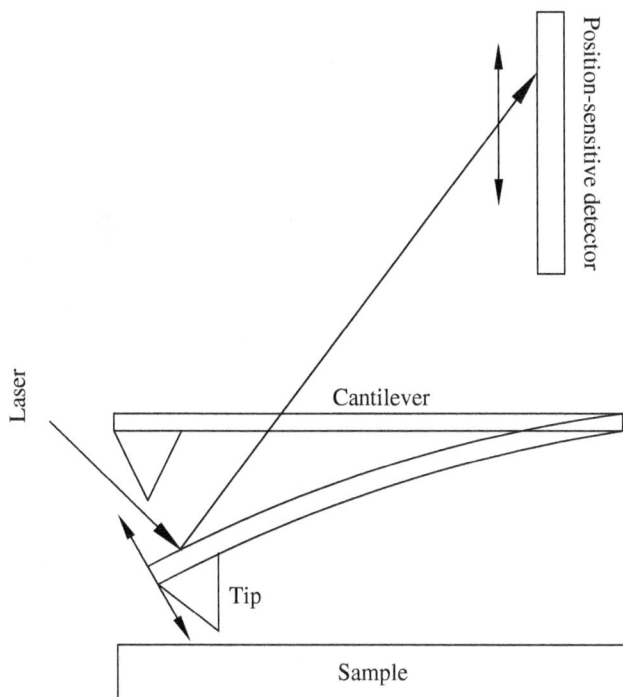

Figure 2.38:   Schematic set-up of an AFM.

the force between tip and sample is approximately linear with the distance. Hence, the deflection of the tip is a measure of the surface topography. An example of an image obtained in contact mode is shown in Figure 2.39. One should also note that AFM images are actually a convolution between the tip shape and the actual surface profile. The sketch shown in Figure 2.40 illustrates the concept and should always be kept in mind.

Resonant modes. Key to understanding resonant modes is the observation that the cantilever is a mechanical oscillator with its own natural frequency of oscillation ω_0. Excitation of the oscillator with an external force can be used to monitor the interaction between cantilever and surface:

$$\ddot{x} + 2\gamma\dot{x} + \omega_0^2 x = A\cos(\omega t) + \frac{f(D,t)}{m}, \qquad (2.49)$$

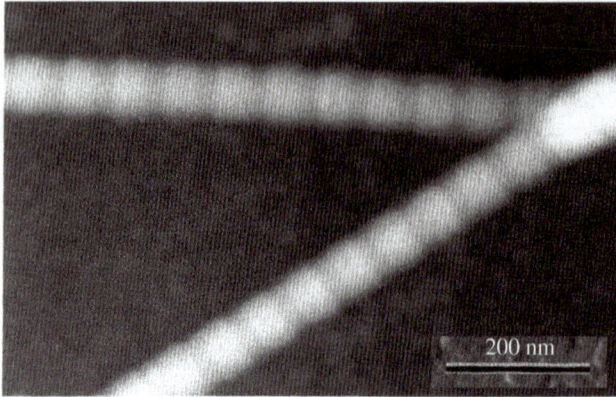

Figure 2.39: AMF image of collagen fibrils. Adapted with permission from Tatsuo *et al* (1996); © International Society of Histology and Cytology.

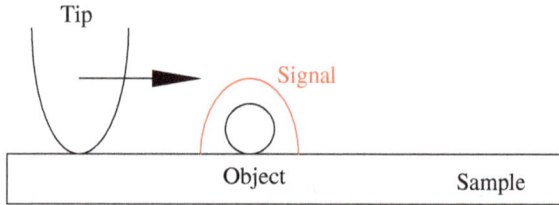

Figure 2.40: Sketch illustrating typical tip artefacts in AFM. The red line indicates the actual image recorded, which depends on the size of the tip and of the feature to be imaged.

where x is the displacement of the tip from its equilibrium position, A is the amplitude of the excitation force, ω its frequency, γ the dissipation term and m the mass of the oscillator. $f(D,t)$ is the force between tip and surface, D being the distance between tip and sample before excitation of the cantilever. The term $f(D,t)$ determines the amplitude, but also the phase, of the oscillation. Over time a series of imaging modes have been developed which make use of the term $f(D,t)$ in one way or the other. These imaging modes allow us to measure (among others) modulus of elasticity, adhesion, friction, magnetization (by using a magnetic tip), electric conductivity (by using a conducting tip) and capacitance (by appyling a modulated voltage) between tip and sample. An overview of imaging methods can be found in Ref. [12].

2.6. Time–of–flight mass spectrometry

A powerful, yet often underestimtated, tool of nanotechnology is mass spectrometry. There are several types of mass spectrometers: magnetic sector mass spectrometers (MSS), quadrupole mass spectrometers (QMS) and time-of-flight mass spectrometers (TOF-MS). MSS and QMS have a comparatively small mass-to-charge limit and can be used to study only relatively small aggregates. For this reason they see limited use in nanotechnology. For example, typical QMS instruments have a maximum mass-to-charge ratio $m/q = 2000$. This means that a QMS can measure a maximum mass of 2000 for singly charged ions, or 4000 for doubly charged ions. Assuming to be interested in aggregates of Au (atomic mass ~ 200), a QMS would be able to access aggregates 10–20 atoms in size. TOF-MS, instead, allows us to work at $m/q \gtrsim 20{,}000$. As we will see in the next chapter, there are a lot of interesting things happening in aggregates with sizes between 30 and 100 atoms, which explains why TOF-MS is preferred to QMS. The working principle of TOF-MS will now be discussed.

In TOF-MS, like in any other mass spectrometer, the analyte needs to be ionized before its mass can be measured. Ionization is achieved in a multitude of ways, the most popular being electrospray ionization, pulsed laser ionization, and electron bombardment ionization. Description of these ionization techniques is beyond the scope of this work. The interested reader can consult mass spectrometry textbooks, such as [13]. In TOF-MS, ions with a charge q are accelerated by a voltage V and then sent through a tube which is kept free of electric fields, as shown in Figure 2.41. At the entrance of the field-free region, the kinetic energy of the ions is equal to the potential energy of the electrostatic field:

$$qV = \frac{1}{2}mv^2 \tag{2.50}$$

In the drift zone (d long) the velocity v of the ions is constant, and $v = d/t$, t being the time that the ions take to fly through the

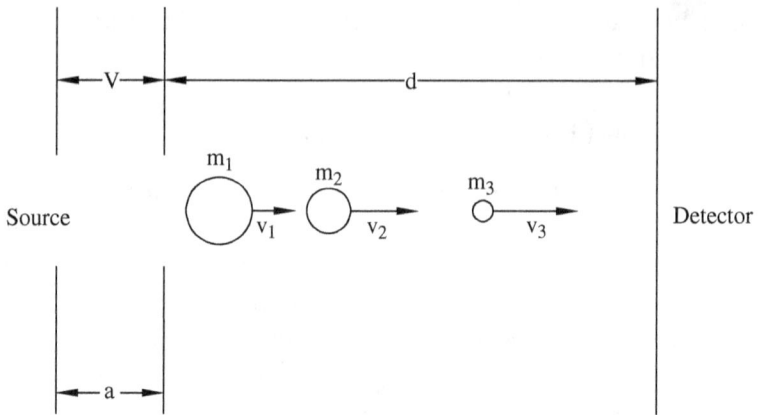

Figure 2.41:   Schematic representation of a TOF-MS.

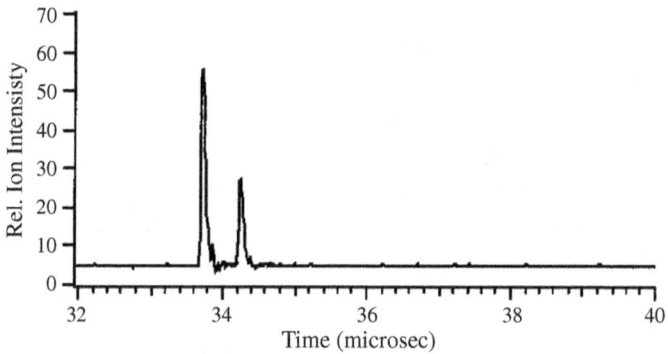

Figure 2.42:   TOF-MS of Cu^+, showing the two isotopes of Cu: ^{63}Cu, 69%, and ^{65}Cu. Adapted with permission from [15]. © American Chemical Society.

drift zone to the detector. Hence:

$$qV = \frac{1}{2}m\left(\frac{d}{t}\right)^2,$$ (2.51)

and therefore:

$$t = d\sqrt{\frac{m}{2qV}}.$$ (2.52)

Using the parameters reported in the seminal work by Wiley and McLaren [14]: $d = 0.40$ m, $V \sim 1$ kV, and a singly-ionized analyte,

Eq. (2.52) yields:

$$t = 2.88 \times 10^{-5} \sqrt{m} \qquad (2.53)$$

These times, on the order of tens of microseconds, are easily measurable. Most importantly, mass differences of a few amu result in time differences of microseconds (that is, measurable without need of sophisticated instrumentation), as shown in Figure 2.42 [15].

References

1. M. Schröder, S. Sallard, M. Böhm, C. Suchomski, B. M. Smarsly, S. Mutisya, M. Einert, M. F. Bertino *Small* **10**, 1566–1574 (2014).
2. B. Williams, D. Clifford, A. A. El-Gendy, E. E. Carpenter, *Journal of Applied Physics* **120**, 033904 (2016).
3. S. Cusack, A. Miller, P. C. J. Krijgsman, J. E. Mellena, *Journal of Molecular Biology* **145**, 525-43 (1981).
4. B. Jacrot, C. Chauvin, J. Witz, *Nature* **266**, 417-21 (1977).
5. D. W. Schaefer, K. D. Keefer, *Phys. Rev. Lett.* **56**, 2199 (1986).
6. G. Beaucage, *J. Appl. Crystallogr.* **29**, 134–146 (1996).
7. L. Reimer, H. Kohl, "transmission Electron Microscopy. Physics of image formation", 5th edition, Sprionger, 2008. ISBN: 978-0-387-40093-8
8. J. Tersoff, D. Hamann, *Phys. Rev. Lett.* **50**, 1998 (1983).
9. S. Kano, T. Tada, Y. Majima, *Chem. Soc. Rev.* **44**, 970 (2015).
10. J. G. Simmons, *J. Appl. Phys.* **34**, 1793–1803 (1963).
11. U. Banin, Y. Cao, D. Katz and O. Millo, *Nature* **400**, 542–544 (1999).
12. C. Dupas, P. Houdy, M. Lahmani (Eds.), "Nanoscience. Nanotechnology and Nanophysics", Springer, Berlin, 2006. ISBN: 3-540-28616-0.
13. J. H. Grossm "Mass Spectrometry: a textbook", Springer, 2011. ISBN: 978 364 210 7092.
14. W. C. Wiley, I. H. McLaren, *Rev. Sci. Instr.* **26**, 1150 (1955).
15. L.-S. Wang, H.-S. Cheng, and J. Fan *J. Chem. Phys.* **102**, 9480 (1995).

Chapter 3

At the Core of Nanotechnology: Clusters of Atoms

Nanotechnology's mission is to investigate the smallest possible atomic aggregates. It turns out that atomic aggregates smaller than about 300 atoms have properties that are often radically different from the bulk. Thus, study of these aggregates is of interest for fundamental research, but also for applications. In the following, different classes of atomic clusters will be described. We will start from the simplest clusters (kept together by dispersion forces) and work our way up to borophene and chalcogenide clusters, which are the most recent additions to cluster science. Please note that in this section we will use the word "cluster" and not "nanocluster". The reason is that cluster science has been around for a very long time. A cluster of atoms indicates an aggregate of a few tens of atoms. It has a dimension of a few nanometers and there is no need to add the "nano" prefix to it.

3.1. Clusters: weakly bound

Here we will focus on clusters of materials that are weakly bound, such as rare gas droplets. This is a mature field, which was started about 40 years ago by a simple necessity: people wanted to understand how these things looked like, and the experimental techniques were still in their infancy. Hence, they started from simple systems.

71

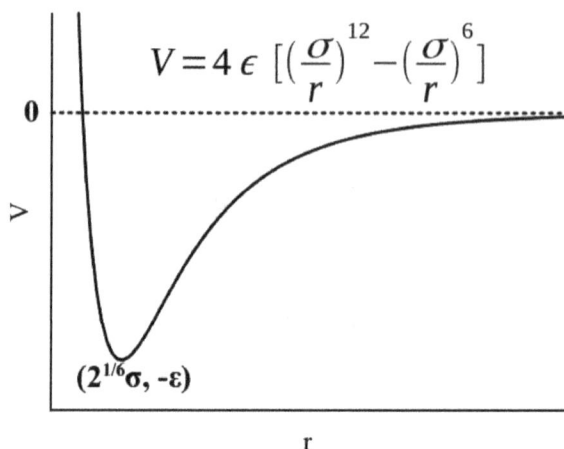

$$V = 4\,\epsilon\left[\left(\frac{\sigma}{r}\right)^{12} - \left(\frac{\sigma}{r}\right)^{6}\right]$$

$(2^{1/6}\sigma, -\varepsilon)$

Figure 3.1: Lennard-Jones potential.

Table 3.1: Binding energies (ε) and melting temperature (T_m) of rare gases and simple molecules.

Gas	ε(kJ/mol)	T_m(K)
Ne	0.4	27
Ar	1.2	87
Xe	2.1	165
H_2	0.3	20
N_2	0.9	77
CH_4	1.3	112

The typical potential that is used to describe this stuff is a Lennard-Jones 6-12, which is shown in Figure 3.1 (r = distance between atoms or molecules).

The Lennard-Jones potential has two fitting parameters (ε and σ) which represent, in essence, the strength of the bond and the interatomic (or intermolecular) distance which minimizes the interaction, as shown in Figure 3.1. The potential is empirical, yet it can reproduce experimental data exceedingly well, without need for complicated and time-consuming *ab initio* calculations. Table 3.1 reports binding energies for selected gases. Two things have be noted. One, the Lennard-Jones potential can be applied not only to noble gases, but also to molecules with a small dipole, such as H_2, and, most importantly, CH_4. Two, the melting point of these gases scales with the binding energy ε. While this may appear trivial, think of the technological implications, such as the liquefaction of methane for the shipping industry and hydrogen storage. For these reasons, it would not be completely surprising to see a resurgence of research on clusters bound by dispersion forces.

3.1.1. *Experimental set up*

To produce rare gas clusters, a molecular beam apparatus is normally used. Such an apparatus consists of three main components. The first component is a nozzle through which the gas expands into vacuum from a pressure of a few atmospheres. This expansion cools the beam to temperatures as low as a few K. These low temperatures enable the study of aggregates of any rare gas, including He. In the simplest implementation the nozzle is followed by a series of collimation apertures and a detector (quadrupole, but, more frequently, a time-of-flight mass spectrometer).

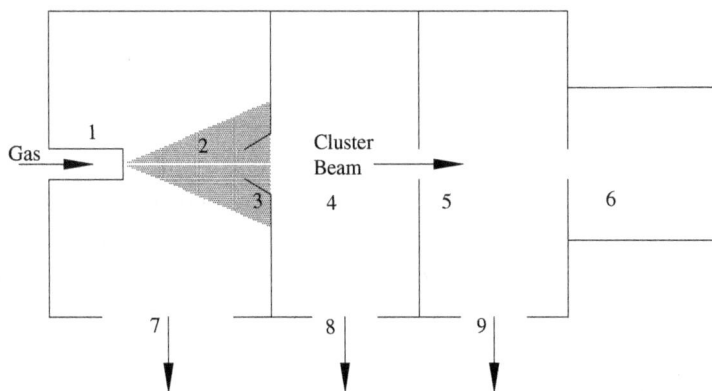

Figure 3.2:   Schematic representation of an apparatus used to produce rare gas cluster beams. 1 = nozzle; 2 = gas, clusters sprayed by the nozzle in all directions. 3 = first collimator (skimmer). 4 = vacuum chamber (with an additional collimating aperture). 5 = Second vacuum chamber, hosting 6 = mass spectrometer. 7,8,9 = vacuum pumps.

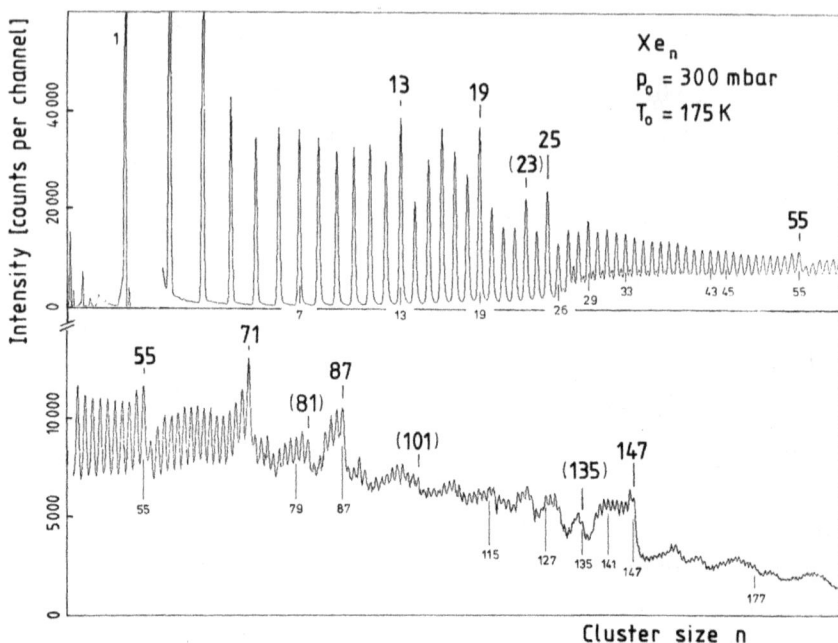

Figure 3.3: Example of mass spectrum of Xe clusters produced by adiabatic expansion through a nozzle. Adapted with permission from [1]. © American Physical Society.

The first experiments of this type were carried with Xe beams. Xe has the highest binding energy among noble gases and it represented the lowest-hanging fruit [1]. A mass spectrum from the seminal paper of Ref. [1] is reported in Figure 3.3. It shows a series of peaks, each of which corresponding to one Xe atom being added to the cluster. The mass spectrum presents two prominent features. The peaks of some clusters (e.g., Xe_{13}) appear to be more intense than their neighbors, and the intensity drops strongly after some numbers (e.g., Xe_{55}).

3.1.2. *Theoretical framework*

Several of these numbers were (and are) well-known to theorists, since they are the number of atoms that can be hosted by platonic solids. Most importantly, platonic solids (in particular, the icosahedron) had been predicted to be the most stable structures by theoretical calculations that preceded experiments [2]. It is instructive

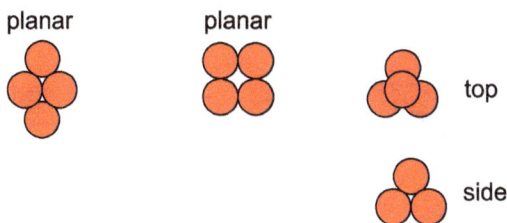

Figure 3.4: Possible configurations of a RG$_4$ cluster (RG = rare gas).

to understand how these structures are calculated theoretically. For N = 1, 2, 3 the solution is trivial. For N = 4 things become a tad more complicated. Atoms could be arranged in a planar geometry (rhombus or square), or in three dimensions as a triangular pyramid, as shown in Figure 3.4. So, how do we decide which is the structure with minimum energy? It is the one where atoms are closest to each other. In this case, the triangular pyramid.

Using the same criteria (minimizing the distance between atoms), one gets the equilibrium geometries of larger clusters. Energy (i.e., distance) minimization is anything but trivial for large structures, but it can be carried out reliably, given the simplicity of the potential. The progression of structures is shown in Figure 3.5, and it shows a preference for the icosahedral motif. Key to the icosahedral motif is the structure of the little-known N = 7 cluster. The most stable structure of this cluster is a pentagonal bipyramid, and the pentagonal base gives rise to icosahedra when the number of atoms in the cluster is increased. Icosahedra, in fact, consist of pentagonal pyramids, as shown in Figure 3.5. The first complete icosahedron is for N = 13, and it corresponds to 12 atoms on the vertexes of the icosahedron, plus one atom in the center. The most stable structures are those where complete icosahedra are formed: N = 13, but also N = 55 (an icosahedron built around an icosahedron), which are shown in Figure 3.5. Going back to Figure 1.12, we note that the most conspicuous features of the spectra are not so much high intensities at the magic numbers, but, rather, intensity drops at magic numbers +1. The intensity drop between 55 and 56 amu is probably the best example of this trend. So, why are intensty drops more prominent than intensity maxima? The reason is shown in Figure 3.6.

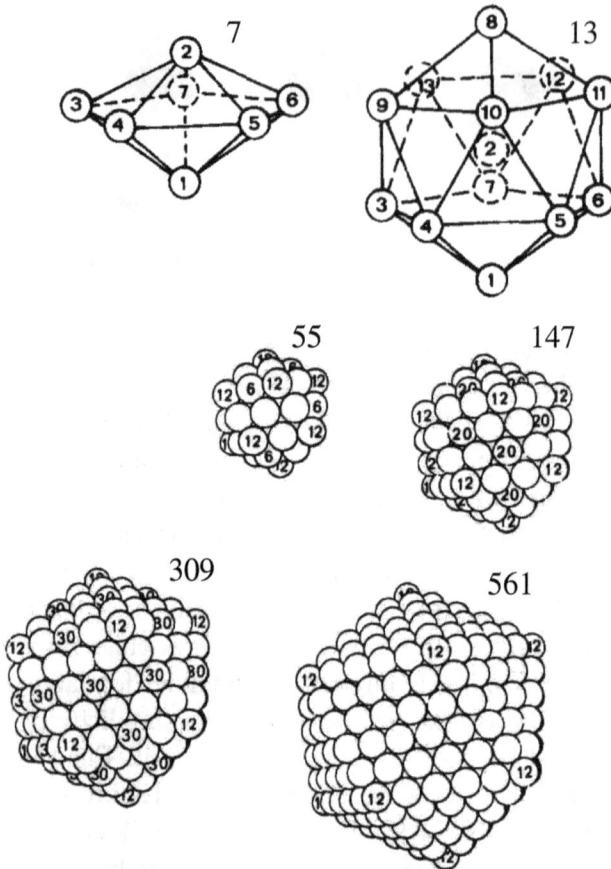

Figure 3.5:   Progression of minimum energy structures of G gas clusters. Adapted with permission from Farges *et al.* (1983) and Anagnostatos (1987); © American Institute of Physics and Elsevier.

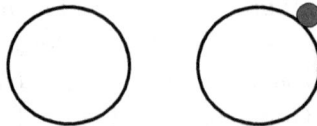

Figure 3.6:   Addition of 1 atom to a magic number cluster (idealized as a sphere) costs a lot of surface energy!

An extra atom added to a magic number cluster will be extremely poorly uncoordinated and stick out like a sore thumb. The energetic price to pay is prohibitive, thus explaining the intensity drop.

3.1.3. *Additional remarks and, as usual, pitfalls*

Clusters with the same number of atoms but a different structure are called isomers. They are quite common, so one should always bear this possibility in mind when working with clusters. Now, the energies of different isomers are often quite comparable. This opens the door (trap door, actually) to artefacts and disputes: theory may suggest wrong structures, and/or experiment may be producing isomers. The situation is made even more complicated by the fact that isomerization may be induced by deposition on a substrate or by addition of a surfactant. A final remark: rare gases are fcc in the bulk, and they do not have the icosahedral structure that they have at the cluster scale. The reason is that atoms on the pentagonal pyramidal cap of an icosahedron are not super-strongly packed. This loose packing costs energy when compared to a fcc structure. In rare gases, the transition to the denser fcc structure occurs between 1,000 and 10,000 atoms.

3.2. Magic numbers in free electron metal clusters

The next logical step from rare gases is the study of free electron metal clusters, such as Na$_N$. Na$_N$ clusters are not very relevant for applications. They would oxidize immediately in atmosphere, and be close to the melting point at room temperature, even when deposited under inert atmosphere. Yet, these clusters represented an important test bed for theory and allowed to develop new experimental techniques.

3.2.1. *Experimental set-up*

A representative experimental set-up for the study of metallic and alloy clusters is shown in Figure 3.7. First, a metal (or carbon, or whatever else one wants to study) is evaporated. There are several ways of achieving evaporation, the most popular being heating a

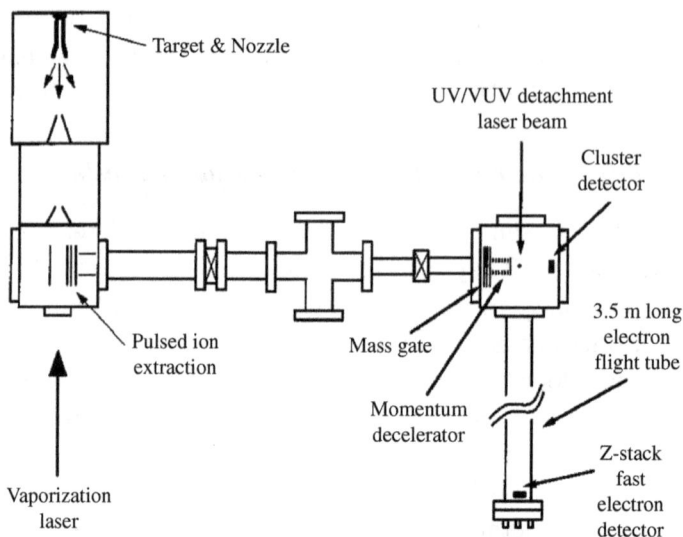

Figure 3.7: Schematic representation of a molecular beam set-up with photoelectron spectroscopy capability. Adapted with permission from [3]. © AIP Publishing.

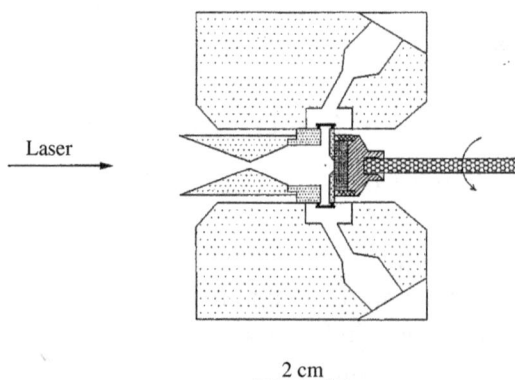

Figure 3.8: Schematic representation of a cluster source. Adapted with permission from [3]. © AIP Publishing.

target with a pulsed laser, as shown in Figure 3.8 [3]. The metal is evaporated in a flow of gas, which is then expanded through a nozzle. Expansion cools down the metal atoms, and collisions force them to coalesce into clusters. After passing through collimating apertures, the negatively charged clusters are extracted perpendicular to the beam using a 1 kV pulse. This pulse starts the clock of the experiment. The ions drift through a TOF-MS spectrometer, 1.5 meters long. A "mass gate" is placed at the end of the TOF arm. This gate is usually at a positive potential (to block the incoming ions). When the time corresponding to the desired mass is reached, the potential is switched off. A detector is placed after the gate, which allows to measure the intensity of the peaks and search for magic numbers if desired. However, something else is possible. The electrons attached to the clusters can be detached using an ultraviolet laser. The energy of the electrons detached from the ion is measured by a second time-of-flight spectrometer. The energy of the photodetached electrons yields information on the electronic states of the cluster.

We will now focus on magic numbers and then extend the discussion to photoelectron spectroscopy.

3.2.2. *The jellium model*

For free electron metals, the theoretical model of choice is the jellium model. In this model, we approximate a cluster with a sphere, positively charged. Electrons will be free to move within this sphere. A simplified potential for this system is $V(\mathbf{r}) = {\infty, r > a \atop 0, r < a}$, that is, the sphere has rigid walls and the potential is constant within the sphere. Given the form of $V(\mathbf{r})$, the Schrödinger equation can be separated into a radial and an angular part, similar to what we do for the hydrogen atom. The radial part of the equation, however, differs from the jellium model and the hydrogen atom:

$$\phi_l'' + \frac{2}{r}\phi_l' + \left(k^2 - l\frac{(l+1)}{r^2}\right) = 0, \quad \text{where } k = \sqrt{\frac{2mE}{\hbar^2}} \quad (3.1)$$

The solution of Eq. (3.1) is the Bessel spherical function:

$$\phi_l = A_l j_l(kr) \quad r \leqslant a$$
$$\phi_l = 0 \quad r > a$$

The boundary condition is $\phi_l = 0$ for $r = a$, that is, $j_l(ka) = 0$.

Therefore: $k_{nl} = \frac{z_{nl}}{a}$, where z_{nl} are the zeros of the Bessel function and n is a positive integer.

The energetic levels are given by:

$$E_{nl} = \frac{\hbar^2}{2ma} z_{nl}^2.$$ (3.2)

The quantum numbers have a familiar flavor. n is the level, and l is the orbital type (s,p,d, ...). However, in the jellium model there is not a limitation as in the hydrogen atom ($l \leq n$). Therefore, one can have 1s, 1d, and even 1f orbitals. Like in the H atom, we can find combinations of orbitals (i.e., number of electrons) that are particularly stable.

The magic numbers of the jellium model are reported in the table below.

Orbital type	No of electrons
1s, 1p	8
1s, 1p, 1d	18
1s, 1p, 1d, 2s	20
1s, 1p, 1d, 2s, 1f	34

For different forms of the potential, and especially when the repulsion between electrons is taken into account, the results cannot be calculated in analytical form, but the general trend reported in the table holds. The first experimental work was carried out on Na and K clusters, metals that are exceedingly easy to evaporate. The results are reported in Figure 3.9 and show magic numbers as predicted by the jellium model.

The jellium model can be extended to nearly-free-electron metals such as Cu, Ag, Au and Al, and an excellent review of the field is reported in Ref. [4]. Of particular relevance are results on the cluster Al_{13}^- [5]. The cluster was found especially resistant to oxidation, which is no surprise. Al has a valence of 3; therefore, in first approximation, an Al_{13} cluster will have 39 "free" electrons. Add one negative charge and a magic number (40, jellium-wise) is reached. For

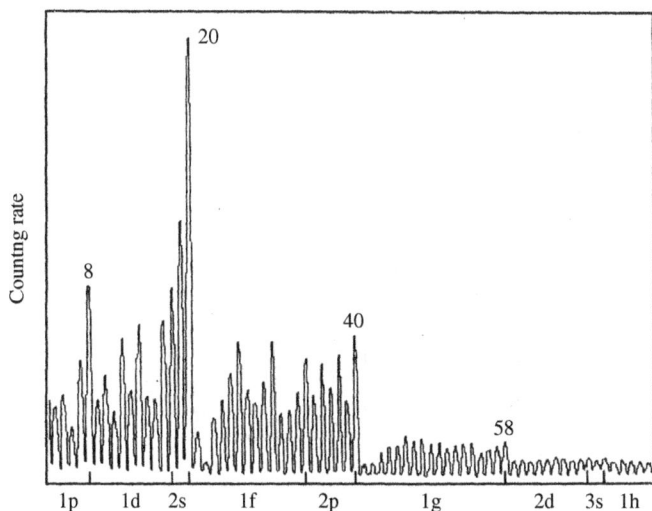

Figure 3.9: Abundance of Na clusters. Note the drop at the positions predicted by the jellium model. Adapted with permission from Clemenger (1985); © American Physical Society.

the same reason, Cu_{20}, Cu_{40}, Cu_{58} and Cu_{92} are also resistant to oxidation [15].

3.3. Beyond the jellium model

While powerful, the jellium model is limited to a few metals and cannot be applied if the bond between the atoms in the clusters is covalent. For these other systems we do not have any simple theoretical framework. We have to rely on *ab initio* calculations, analogies with inorganic and organometallic chemistry, and sometimes sheer luck, in the hope of finding stable configurations. The most blatant case is that of C_{60}, which had not been predicted by any theory but was found to be prominent in laser vaporization spectra, as shown in Figure 3.10 [6].

Another prominent example is that of metal carbides (Met-Cars) with the general formula $M_8C_{12}^+$, where M = Ti, V, Hf, Nb, Zr; an example of which is shown in Figure 3.11 [7].

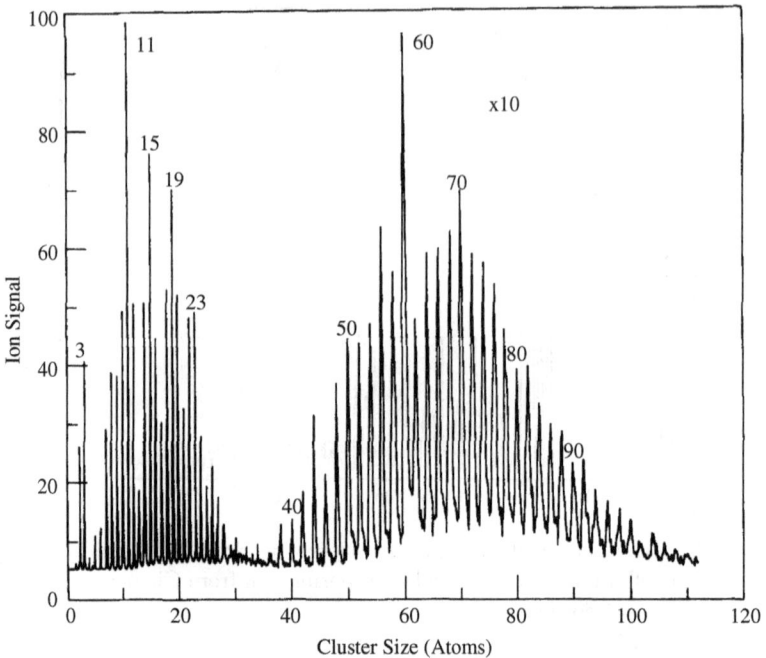

Figure 3.10: Mass spectra of carbon clusters. Adapted with permission from [6]. © AIP Publishing.

None of these structures (C_{60}, Met-Cars) had been predicted as being particularly stable. However, once experiment determined their stability, calculations did indeed show that this stuff was energetically favored.

With beam cluster experiments, one can determine not only mass abundances, but also physical properties such as ionization potentials, electronic structure and magnetism. These experiments help understand the structure of the clusters, but also the differences from the bulk and the evolution from cluster to bulk electronic properties. We will now discuss two examples, based on photoelectron spectroscopy. The first example, for Cu_n^- clusters, is reported in Figure 3.12 [8]. The smallest clusters show binding energies that are close to those of Cu^- anions. No big surprise here. When the cluster size increases, a band around 6 eV becomes more prominent. The energy of this band increases with cluster size and gets within 0.6 eV of the

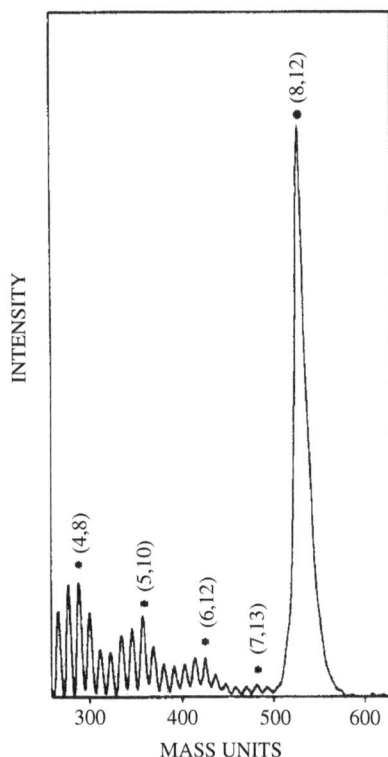

Figure 3.11: Met-Cars obtained by reaction Ti^+ ions with CH4. Note the prominence of the $Ti_8C_{12}^+$ peak. Adapted with permission from [7]. © American Chemical Society.

bulk work function for the largest clusters. This is also not totally surprising. What is noteworthy, though, is that a near-bulk behavior is already developed for cluster sizes on the order of 100 atoms. In terms of size, this means that nanoparticles as small as ~2 nm can be assumed to have an electronic structure close to that of the bulk. The second example is shown in Figure 3.13 for Au_n^- clusters [9]. The experimental spectra show several peaks, which do not present much of a trend. Theoretical calculations show that the spectra can be well reproduced by cage-like structures. Non-hollow structures yield spectra that are not consistent with the experimental data. The experiments of Ref. [9] show the necessity of theoretical calculations,

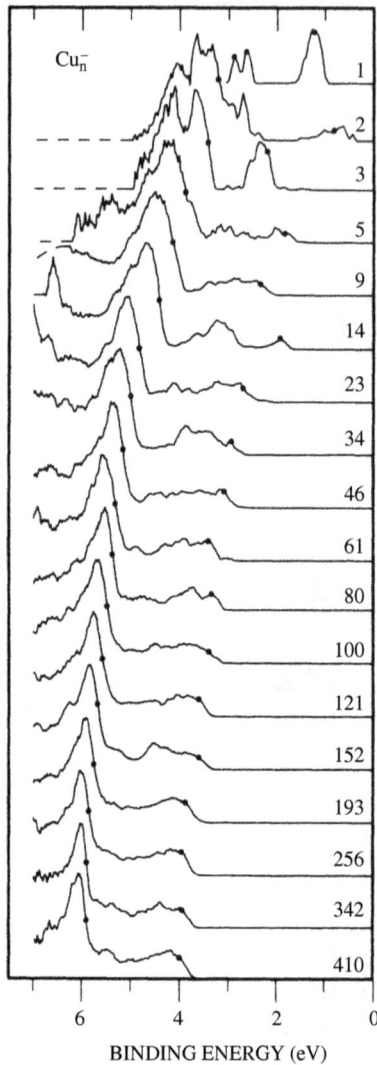

Figure 3.12: Photoelectron spectroscopy of Cu_n^- clusters. Adapted with permission from [8]. © American Physical Society.

without which the experimental data could not be interpreted at all. One must also emphasize that the calculations are anything but simple. To be confident of the simulations, tens of different structures must be analyzed. The process is time-consuming and not trivial.

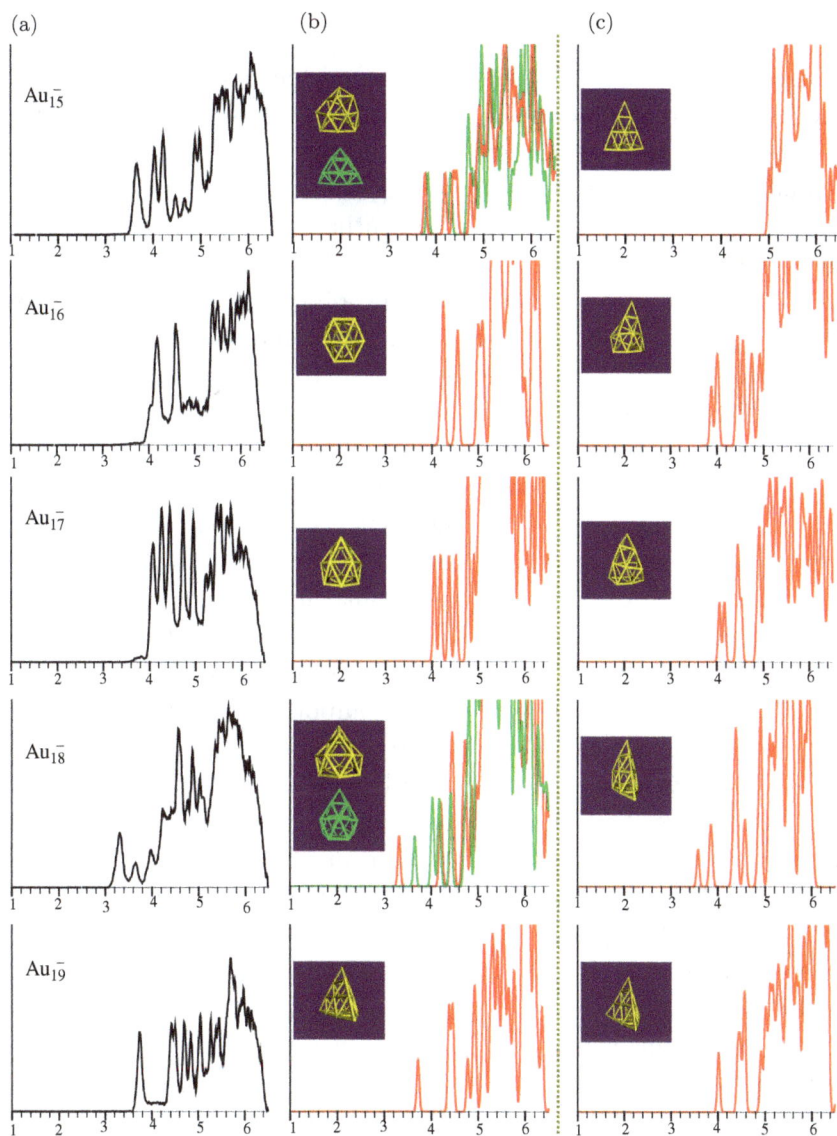

Figure 3.13: (a) Experimental photoelectron spectra of Au_n. (b) Simulated spectra for one (or two) lowest-lying isomers matching the first and second major peaks of the measured spectra. (c) Simulated spectra for the non-hollow-cage candidate isomer, which appears not to match the observed spectra. Adapted with permission from [9]. © National Academy of Sciences.

3.4. Towards cluster-assembled materials

Cluster matter, that is, solids where atomic clusters are the building blocks, has great potential for applications. However, bridging the gap between gas-phase experiments and real-world applications is a daunting task. For example, following the discovery of C_{60}, a number of methods were developed to synthesize it using wet chemical methods. Using solid state chemistry techniques, alkali and transition metal ions were inserted inside the hollow space of the C_{60} cages. However, none of the promised applications materialized. This does not mean that cluster matter is doomed to irrelevance. Failure, or applications in fields far from what was originally expected, is the rule in materials development. For example, C_{60} was proposed and apparently employed for some time in skin care products.

Current discovery and upscaling efforts focus on two strategies. One strategy is to take a page from nature and try to replicate materials that are known to have cluster-like units. For example, it is known that the reaction of metals such as Pb, Sn, Si or Ge with alkali metals (except for Li) yields solids with tetrahedral building blocks (Zintl phases), and that some Boron compounds present an icosahedral motif. Correspondingly, wet- and solid state- chemical techniques have been developed for the synthesis of these compounds; some of which (boranes) have found applications in medicine [10]. Here, we will focus on recent and extremely promising developments, which are reported in Ref. [11]. In this work, two types of cluster units were synthesized. One type of clusters, found in (macroscopic) Chevrel phases, has the formula M_6E_8, where M = metal (for example, Co) and E = chalcogenide (for example, Te). The second type of clusters, found in proteins containing Fe and S, has the formula M_4E_4. The structures of these clusters are reported in Figure 3.15. These clusters can donate charge to C_{60} and form ordered crystals, as shown in Figure 3.14. These ordered crystals are the equivalent of, say, NaCl. Only, the building blocks are not atoms, but clusters! These clusters can also self-assemble into thin films, and in millimeter-sized crystals which resemble layered solids such as mica. More recently, the same group demonstrated that thin films of the compound $Re_6Se_8Cl_2$, also

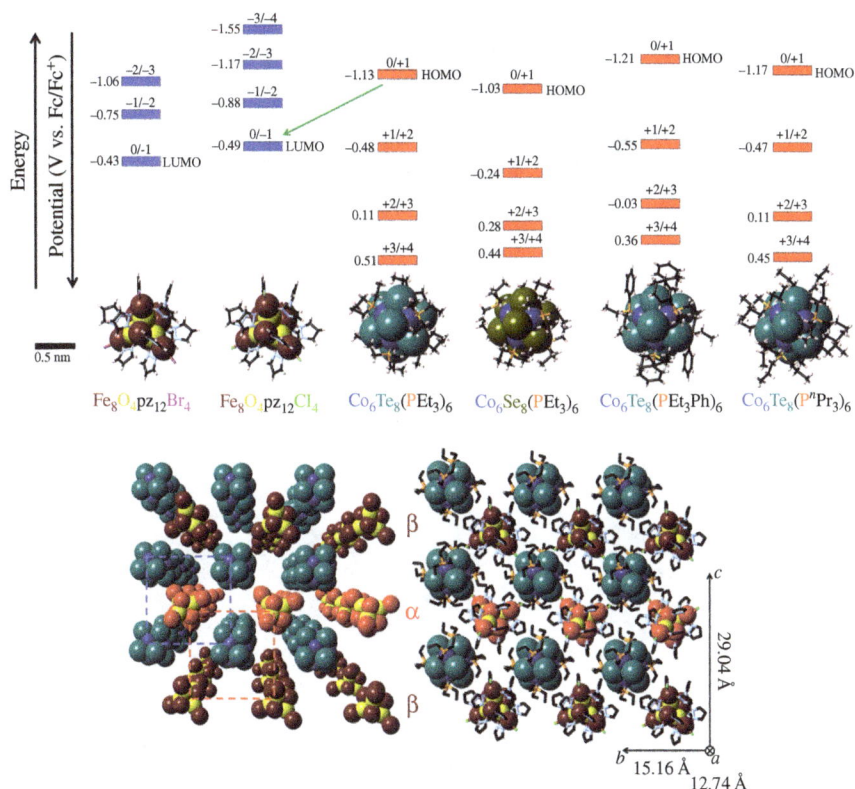

Figure 3.14: Molecular structure, energetic levels and crystal structure of M_8E_8, M_8E_4 building blocks. Adapted with permission from [11]. © American Chemical Society.

found in Chevrel phases, are semiconductors with a band gap of 1.49 eV. Doping (n-type) can be achieved by removal of the chlorine atoms [12].

The second strategy is to look at clusters with a planar structure, which can be used as building blocks of thin films. The best example is that of boron clusters. Boron clusters were first analyzed by mass spectrometry, and no magic numbers could be identified [13]. The spectrum is unremarkable, without any evidence of magic numbers. The interesting part comes from photoelectron spectroscopy experiments, which are best simulated by assuming planar structures. The structure and molecular orbitals of these boron

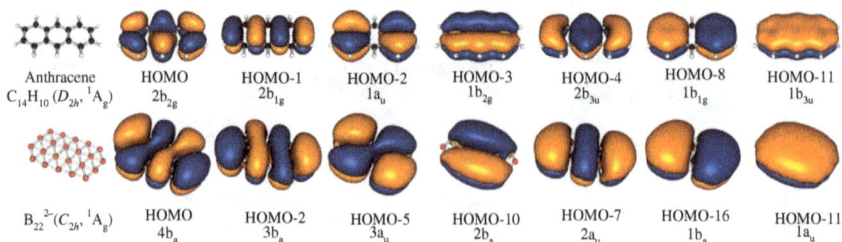

Anthracene $C_{14}H_{10}$ (D_{2h}, ^1A_g) HOMO $2b_{2g}$ HOMO-1 $2b_{1g}$ HOMO-2 $1a_u$ HOMO-3 $1b_{2g}$ HOMO-4 $2b_{3u}$ HOMO-8 $1b_{1g}$ HOMO-11 $1b_{3u}$

B_{22}^{2-}(C_{2h}, ^1A_g) HOMO $4b_a$ HOMO-2 $3b_a$ HOMO-5 $3a_u$ HOMO-10 $2b_a$ HOMO-7 $2a_u$ HOMO-16 $1b_a$ HOMO-11 $1a_u$

Figure 3.15: Comparison between the structure and molecular orbitals of anthracene (top) and of the best fit of B_{22}^{2-} to photoelectron spectroscopy data. Adapted with permission from Sergeeva *et al.* (2012); © American Chemical Society.

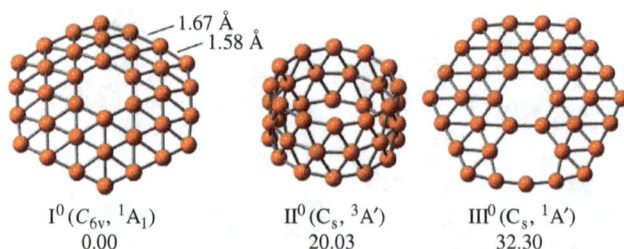

I^0 (C_{6v}, $^1A_1$) 0.00 II^0 (C_s, $^3A'$) 20.03 III^0 (C_s, $^1A'$) 32.30

Figure 3.16: Calculated structure of B_{36}. Adapted with permission from Z.A. Piazza *et al.* (2013); © MacMillan Publishers Limited.

clusters present uncanny analogies with aromatic compounds, as shown in Figure 3.15. These clusters have been proposed as non-carbon analogues of aromatic compounds. For thin film applications, the clusters B_{35} and B_{36} play a particularly important role. Their structure is nearly planar, and they present a central vacancy, as shown in Figure 3.16. This structure is very relevant, because theoretical calculations had predicted that boron could form 2D structures (borophenes) analogous to graphene, provided that vacancies could be created in the center of hexagonal structures [14]. Evaporation of boron on Ag substrates under ultra high vacuum conditions (to prevent contamination) yielded the much-looked-for 2D structures, which are shown in Figure 3.17 [15]. Given the novelty of these results, it is still too early to predict if these systems will have any novel properties and, most importantly, if we will ever

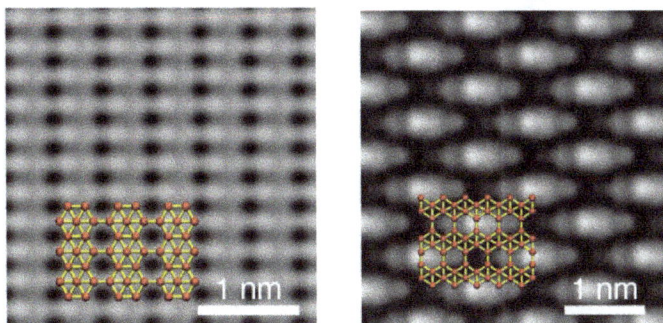

Figure 3.17: STM images of boron thin films deposited on Ag(111). The structures calculated from the imags are shown in color. Adapted with permission from [15]. © Springer Nature.

be able to upscale the production and make devices out of them. Yet, the research is extremely promising and surely worthy of being pursued.

References

1. O. Echt, K. Sattler, and E. Rechnagel, *Phys. Rev. Lett.* **47**, 1121 (1981).
2. M. R. Hoare and P. Pal, *Nature Physical Science* **230**, 5–8 (1971).
3. L.-S. Wang, H.-S. Cheng, J. Fan, *J. Chem. Phys.* **102**, 9480 (1995).
4. J. A. Alonso, "Structure and properties of atomic nanoclusters", Imperial College Press, London (2005).
5. R. E. Leuchtner, A. C. Harms, and A. W. Jr. Castleman, *J. Chem. Phys.* **94**, 1093–1101 (1991).
6. E. A. Rohlfing, D. M. Cox, A. Kaldor, *J. Chem. Phys.* **81**, 3322 (1984); see also H. W. Kroto, A. W. Allaf, S. P. Balm, *Chem. Rev.* **91**, 1213–1235 (1991).
7. S. Wei, B. C. Guo, J. Purnell, S. Buzza, A. W. Castleman, Jr. *J. Phys. Chem* **96**, 4166–4168 (1992).
8. O. Cheshnovsky, K. J. Taylor, J. Conceicao, R. E. Smalley, *Phys. Rev. Lett.* **64**, 1785 (1990).
9. S. Bulusu, X. Li, L.-S. Wang, and X. C. Zeng, *Proceedings of the National Academy of Sciences*, **103**, 8326–8330 (2006).
10. K. Shelly, D. A. Feakes, M. F. Hawthorne, P. G. Schmidt, T. A. Krisch, W. F. Bauer, *Proceedings of the National Academy of Sciences of the United States of America*, **89**, 9039–9043 (1992).
11. A. Turkiewicz, D. W. Paley, T. Besara, G. Elbaz, A. Pinkard, T. Siegrist, X. Roy, *J. Am. Chem. Soc.* **136**, 45, 15873–15876 (2014).

12. X. Zhong, K. Lee, B. Choi, D. Meggiolaro, F. Liu, C. Nuckolls, A. Pasupathy, F. De Angelis, P. Batail, X. Roy, and X. Zhu, *Nano Lett.* **18**, 1483–1488 (2018).
13. L. S. Wang, *International Reviews in Physical Chemistry* **35**, 69–142 (2016).
14. H. Tang and S. Ismail-Beigi, *Phys. Rev. Lett.* **99**, 115501 (2007).
15. A. J. Mannix, Z. Zhang, N. P. Guisinger, B. I. Yakobson, M. C. Hersam, *Nature Nanotechnology* **13**, 444–450 (2018).

Chapter 4

Plasmonics

4.1. Theory of mechanical resonance

Plasmons are resonances. To understand them, it is best to refresh our memory of mechanical resonance.

Let us consider a harmonic oscillator, that is, a spring of elastic constant k_{el} to which we connect a mass m, as shown in Figure 4.1. If we now stretch the spring by a distance x from its equilibrium position, we end up with a recalling force $F = -k_{el}x = m\frac{d^2x}{dx^2}$ which leads to oscillations with frequency $\omega_0 = \sqrt{\frac{k_{el}}{m}}$. In the real world, there is damping. In its simplest form, the damping term has the form $c\frac{dx}{dt}$, that is, it is proportional to the speed of the object. In real-world applications a periodic force is also applied to the system. Think of a person pushing someone on a swing set, for example. This force will have the form $F(t) = F_0\sin(\omega t)$. The equation of motion of the mass becomes $m\frac{d^2x}{dx^2} - c\frac{dx}{dt} + kx = F(t)$. The solution is a sinusoidal function with an amplitude that depends on the frequency of the applied function F: $x(t) = A(\omega)\sin(\omega't)$, and the relevant part is $A(\omega)$:

$$A(\omega) = \frac{F_0}{m\omega\sqrt{c^2 + \frac{(\omega_0^2-\omega^2)^2}{\omega^2}}}. \tag{4.1}$$

Let us now analyze Eq. (4.1). If damping is negligible, c = 0 and therefore $A(\omega) = \frac{F_0}{m(\omega_0^2-\omega^2)}$. This means that A becomes very large when $\omega \sim \omega_0$. This is the basic idea of resonance. That is, the system has a "natural" frequency of oscillation $\omega_0 = \sqrt{\frac{k_{el}}{m}}$ when left to its own devices. When a periodic force is applied with a frequency close to the natural frequency, the amplitude of the oscillations keeps increasing, much in the same way as in a swing set. Damping is the additional term in the denominator of Eq. (4.1) which prevents the denominator to go to zero and limits the maximum amplitude of the oscillations. Also, and let us remember this, damping increases the width of the resonance, as shown in Figure 4.2.

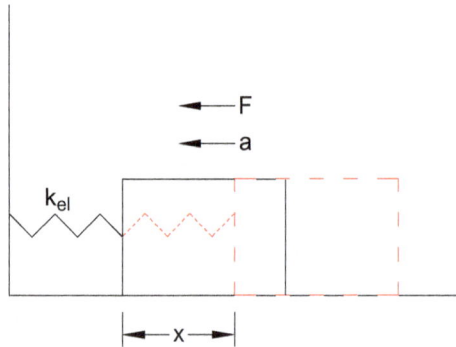

Figure 4.1:   Schematic representation of a mechanical harmonic oscillator.

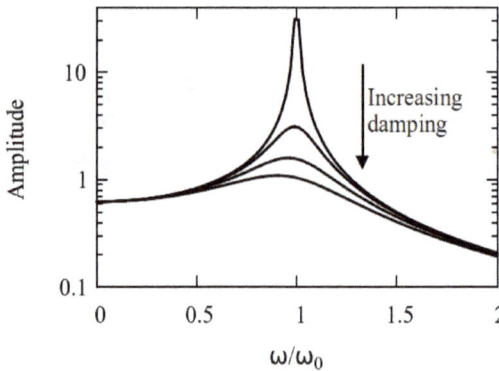

Figure 4.2:   Amplitude of periodic oscillations near the resonance conditions. Note how the resonance broadens with increasing damping.

4.2. Resonances in free electron metals

We now move to metals, which we approximate as positive cores at fixed positions, with electrons that can move freely around the cores. In the ground state, the electrons will be close to their cores, as shown in Figure 4.3.

Assume now to displace the electron cloud by a distance d from the equilibrium position, as shown in Figure 4.4. The displacement of the charges creates a dipole and an electric field that tends to recall the electron cloud back to its equilibrium position. The hand-waving explanation of what happens next is as follows. The recalling electrical field is the analog of the spring in a mechanical oscillator and generates oscillations with a natural frequency ω_0. An electromagnetic (e.m.) field interacting with the metal plays the same role of

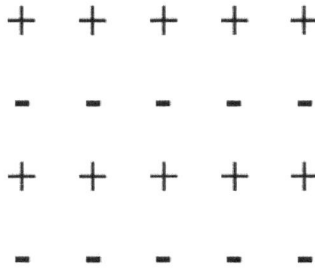

Figure 4.3: Idealilzed model of a free electron at rest: The electrons are close to the fixed, positively charged, atomic cores.

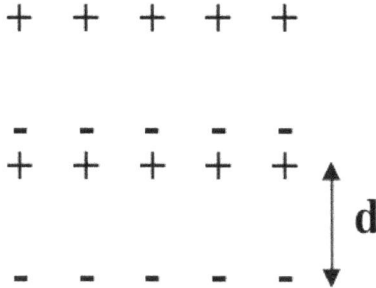

Figure 4.4: Displacement of the electron cloud by a distance d from its equilibrium position.

the periodic force in the mechanical oscillator. When the frequency of the e.m. field is close to that of the natural frequency, a resonance occurs and the e.m. field is absorbed by the material. This is the idea behind plasmons.

A more rigorous derivation of the plasmonic frequency is as follows. We assume that the volume density of the electrons is N, and that the electronic cloud is displaced by a distance $d = x$. The surface charge is then $\sigma = Nxe$ and the magnitude of the recalling field is $E = 4\pi\sigma = 4\pi Nxe$. Thus, $m\frac{d^2x}{dt^2} = -eE = -4\pi e^2 x$. This is the equation of a harmonic oscillator with frequency:

$$\omega_p^2 = \frac{4\pi Ne^2}{m}. \tag{4.2}$$

We call the frequency ω_p the plasma frequency.

4.3. Elements of solid state optics

Before we get more in depth into plasmonic effects, we need to discuss the interaction of electromagnetic fields with solids. First, we start from Maxwell's equations, which we have seen already.

$$\nabla \cdot \mathbf{D} = 0 \quad \nabla \cdot \mathbf{B} = 0$$
$$\nabla \times \mathbf{E} = \frac{-1}{c}\frac{\partial \mathbf{B}}{\partial t} \nabla \times \mathbf{B} = \frac{4\pi}{c}\mathbf{J} + \frac{1}{c}\frac{\partial \mathbf{D}}{\partial t} \tag{4.3}$$
$$\mathbf{D} = \mathbf{E} + 4\pi\mathbf{P} \quad \mathbf{B} = \mathbf{H} + 4\pi\mathbf{M}.$$

We also assume to work with non-magnetic materials, that is, $M = 0$ and $\mathbf{B} = \mathbf{H}$.

We also assume that the response of the material to the external fields is linear:

$$\mathbf{P} = \alpha\mathbf{E}$$

$$\mathbf{J} = \sigma\mathbf{E}, \tag{4.4}$$

where α is the polarizability and σ the conductivity of the material.

Under these assumptions:

$$\mathbf{D} = \mathbf{E} + 4\pi\mathbf{P} = (1 + 4\pi\alpha\mathbf{E}) = \epsilon\mathbf{E},$$

that is,

$$\epsilon = 1 + 4\pi\alpha \tag{4.5}$$

We also assume the material to be isotropic, which means that ε and α are scalars, not tensors.

We now let an electric field interact with the material. The field will be of the form $\mathbf{E} = \mathbf{E_0}e^{-i\omega t}$. Feeding this field into Maxwell's equations, we obtain:

$$\nabla \times \mathbf{B} = \frac{1}{c}\frac{\partial(\epsilon\mathbf{E})}{\partial t} + \frac{4\pi}{c}(\sigma\mathbf{E}) = \frac{1}{c}\left(\epsilon\frac{\partial\mathbf{E}}{\partial t} - 4\pi\sigma\mathbf{E}\right) = \frac{1}{c}\tilde{\epsilon}\frac{\partial\mathbf{E}}{\partial t}, \tag{4.6}$$

where $\tilde{\epsilon}$ is the **COMPLEX** dielectric function:

$$\tilde{\epsilon} = \epsilon + \frac{i4\pi\sigma}{\omega} = \epsilon_1 + i\epsilon_2. \tag{4.7}$$

We can now derive an equation for e.m. wave propagation inside the material. We pretty much proceed as one does for the derivation of the wave equation in vacuum, with the only difference that we use $\tilde{\epsilon}$ of the material and not of vacuum. At the end of the day, we obtain:

$$\nabla^2\mathbf{E} = \frac{\tilde{\epsilon}}{c^2}\frac{\partial^2\mathbf{E}}{\partial t^2}. \tag{4.8}$$

As per usual, the solution of Eq. (4.8) has the form

$$\mathbf{E} - \mathbf{E_0}e^{i(k\cdot r - \omega t)} + complex\ conjugate, \tag{4.9}$$

the complex conjugate being a technicality needed to keep $\mathbf{E}$ a real number. Substituting Eq. (4.9) into Eq. (4.8) we get a fundamental relation, which is called the dispersion relation:

$$\omega^2 = \frac{c^2}{\tilde{\epsilon}^2} k^2. \tag{4.10}$$

From this relation we get that $k = \frac{\omega\tilde{\epsilon}}{c}$ so that Eq. (4.9) becomes

$$\mathbf{E} = \mathbf{E_0} e^{-i\omega(t - \frac{\tilde{\epsilon}x}{c})} + complex\ conjugate \tag{4.11}$$

Now, remember from Eq. (4.7) that $\tilde{\epsilon}$ is complex. Substituting Eq. (4.7) into Eq. (4.11):

$$\mathbf{E} = \mathbf{E_0} e^{\frac{\omega x \epsilon_2}{c}} e^{-i\omega(t - \frac{\epsilon_1 x}{c})} + complex\ conjugate, \tag{4.12}$$

which shows that in one direction the field is attenuated (in this case, the negative x direction). This means that the material will absorb part of the incident radiation if $\varepsilon_2 \neq 0$.

4.4. Putting it together: Dielectric function, absorption and plasma frequency

Let us now assume that single electrons behave as linear oscillators with a natural frequency ω_p, which is the basic assumption of the Thomson model. This approximation is crude but its results are instructive. We also assume that the motion of the electrons is damped and that damping is proportional to the velocity.

The equation of motion of the electrons is:

$$m\frac{d^2x}{dt^2} + m\gamma\frac{dx}{dt} + \omega_0^2 mx = eE_0 e^{-i\omega t}, \tag{4.13}$$

and the solution is:

$$x = \frac{eE_0 e^{-i\omega t}}{m[(\omega_0^2 - \omega^2) - i\gamma\omega]}. \tag{4.14}$$

We know from the theory of electromagnetism that

$$J = \sigma E = Nev = Ne\frac{dx}{dt}, \tag{4.15}$$

where N is the electron density.

Thus:

$$\sigma = \frac{Ne}{E}\frac{dx}{dt} = \frac{Ne^2(-i\omega)E}{Em[(\omega_0^2 - \omega^2) - i\gamma\omega]}. \tag{4.16}$$

Also, the polarizability P can be roughly expressed as the displacement of the electron from its equilibrium position, induced by the field E:

$$P = Nex = \tilde{\alpha}E, \tag{4.17}$$

therefore:

$$\tilde{\alpha} = \frac{Nex}{E} = \frac{Ne^2}{m[(\omega_0^2 - \omega^2) - i\gamma\omega]}. \tag{4.18}$$

Equation (4.5) becomes:

$$\tilde{\epsilon} = 1 + 4\pi\tilde{\alpha} = \epsilon_1 + i\epsilon_2, \tag{4.19}$$

and we obtain:

$$\epsilon_1(\omega) = 1 + \frac{4\pi e^2 N}{m}\frac{\omega_0^2 - \omega^2}{(\omega_0^2 - \omega^2)^2 + \omega^2\gamma^2}$$

$$\epsilon_2(\omega) = \frac{4\pi e^2 N}{m}\frac{\omega\gamma}{(\omega_0^2 - \omega^2)^2 + \omega^2\gamma^2}. \tag{4.20}$$

In free electron systems, $\omega_0 = 0$, since electrons are not bound to single atoms. Using Eq. (4.2), we get:

$$\epsilon_1(\omega) = 1 - \frac{\omega_p^2}{\omega^2 + \gamma^2}$$

$$\epsilon_2(\omega) = \frac{\omega_p^2\gamma}{\omega^3 + \omega\gamma^2}. \tag{4.21}$$

Equation (4.21) tells us that in free electron metals the dielectric constant has an imaginary component. We also note that in Eq. (4.21) there is a damping term, γ. This damping term is related to the time τ between collisions of an electron with the fixed ions of the metal: $\gamma = \frac{1}{\tau}$. Physically, short τ indicates that the material has a high electrical resistivity. For most metals: $\tau \approx 10^{-14}s$, therefore $\gamma \sim 10^{14}$ Hz. This is a very high frequency. So, for most applications, $\omega \ll \frac{1}{\tau}$, that is, ω is much smaller than γ. Then, Eq. (4.21) becomes:

$$\epsilon_1(\omega) \approx 1$$

$$\epsilon_2(\omega) \gg 1, \tag{4.22}$$

which means that the system will strongly absorb e.m. radiation. When, instead, $\omega > \frac{1}{\tau}$:

$$\epsilon_1(\omega) \approx 1 - \frac{\omega_p^2}{\omega^2} \to 1 \quad \text{when } \omega \gg \omega_p$$

$$\epsilon_2(\omega) \approx \frac{\omega_p^2 \gamma}{\omega^3} \to 0 \qquad \text{when } \omega \gg \omega_p \tag{4.23}$$

which means that the material becomes transparent at high frequencies.

4.5. Plasmons in nanoparticles

What we saw until now is for bulk optical properties. The next question is what happens when an e.m. wave interacts with a nanoparticle. The nanoparticle has a dielectric constant ϵ and is surrounded by a medium with dielectric constant ϵ_m. This is the starting point of Mie's theory, which (IMPORTANT!) was derived in 1905 for a generic nanoparticle, not necessarily for a metallic one. The geometry of Mie's theory is shown in Figure 4.5. A spherical particle interacts with a field, and the scattered intensity is analyzed as a function of the angle θ. The bottom line is that the e.m. field interacts with the dipole that is created at the surface of the sphere, pretty much the same way that we saw earlier for a thin film. Only, the spherical geometry makes the mathematical formulation quite complex. Given the complexity of the mathematical formalism, we will now focus on limiting cases, for which analytical trends can be derived [1].

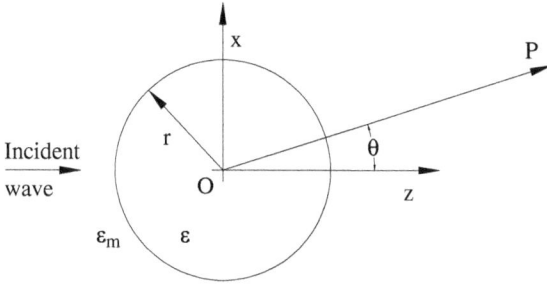

Figure 4.5: Geometry of Mie scattering.

4.5.1. *Small, electrically conducting spheres*

For electrically conducting spheres with a radius r such that:

$$x = \frac{2\pi r}{\lambda} \leq 0.6,$$

an extinction cross section can be calculated [1]:

$$\sigma_{ext} = x^4 \mathrm{Re}\left(\frac{8}{3}\left(\frac{n^2-1}{n^2+2}\right)\right)$$
$$- \mathrm{Im}\left(4x\frac{n^2-1}{n^2+2} + \frac{4}{15}x^3\left(\frac{n^2-1}{n^2+2}\right)\frac{n^4+27n^3+38}{2n^2+3}\right). \quad (4.24)$$

In Eq. (4.24), the first term is the familiar Rayleigh scattering. This term exhibits a λ^{-4} dependence and it is responsible for the preferential scattering of blue light by nanoparticle suspensions. The second term is an absorption term and has a complex dependence on the index of refraction n. There are a couple of things to remember here: $n = \sqrt{\epsilon}$, ε is a complex number, and $\varepsilon = \varepsilon(\lambda)$, that is, the dielectric function depends on the wavelength. In addition, λ refers to the wavelength of the incident wave *inside* the material. That is, if the incident wave has a vacuum wavelength λ_v, the wavelength in the material will be $\lambda = \lambda_v/n$, where n is the index of refraction of the material where the wave is propagating.

If the particles are really small ($r \ll \lambda$), the scattering term is negligible [1] and

$$\sigma_{ext} = constant \times \frac{\epsilon_2(\omega)}{\left(\epsilon_1(\omega) + 2\epsilon_m(\omega)\right)^2 + \epsilon_2^2(\omega)}. \tag{4.25}$$

In Eq. (4.25), we notice a resemblance with the equation for the mechanical resonance amplitude, Eq. (4.1). That is, there is a resonance condition:

$$\epsilon_1 + 2\epsilon_m = 0, \tag{4.26}$$

as well as a damping term (ε_2). Using the dielectric constant for the free electron model, Eq. (4.23), in Eq. (4.26), we obtain for the resonance frequency:

$$\omega_{resonance} = \frac{\omega_p}{\sqrt{(2\epsilon_m + 1)}} = 2e\sqrt{\frac{\pi N}{(2\epsilon_m + 1)m}}. \tag{4.27}$$

Let us now look at Eq. (4.27) more in detail. Its shape is very similar to Eq. (4.2), and indicates that at the resonance condition the electron cloud will bounce back and forth and behave like in the simplified model of Figure 4.4. Eq. (4.27) also tells us that the resonance frequency will depend on the electron density N and on the dielectric constant of the medium where the nanoparticles are embedded (ε_m). In air, where $\varepsilon_m = 1$, the resonance condition, Eq. (4.27) becomes:

$$\omega_{nanoparticle,air} = \frac{\omega_p}{\sqrt{3}}. \tag{4.28}$$

We also note that Eq. (4.25) has a damping term, ε_2. Therefore, whenever a metal has a large ε_2 at the resonance condition, the resonance will be weak and broad. Figure 4.6 shows that the width of the plasmon resonance is considerably different for Au, Ag and Cu, EXACTLY because of the differences in ε_2 [2]. This example explains why most metals (with the exception of the alkalis and the coinage metals) do not exhibit a resonance: ε_2 is very large for most metals over a wide range of frequencies.

At this point, we have to be careful with words. The resonance condition in metal nanoparticles, Eq. (4.27), is often called "surface plasmon". The term "surface" comes from the fact that the restoring

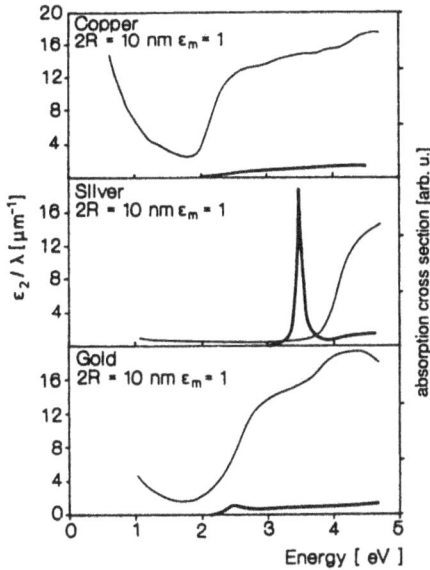

Figure 4.6: Absorption cross section (thick black line) and ε_2 (thin line) of the coinage metals as a function of incident energy. Note how the intensity of the resonance decreases and its width increases with increasing ε_2 at the resonance position. Adapted with permission from [2]. © Springer Verlag.

force comes from the surface polarization, the same way as in the thin film/bulk case. Also, one can show that the amplitude of the resonance is about one atom, hence the "surface" term. We also note that what we call here "surface plasmon resonance" is called by some researchers "localized surface plasmon resonance". The reason for the name is to distinguish the plasmonic resonance in thin films (which will be described in the following) from the plasmon resonance in nanoparticles.

4.5.2. *Size dependence*

Equation (4.24) shows that the extinction coefficient depends on particle size via the variable x. However, for many systems the position and shape of the surface plasmon do not depend strongly on particle size. Figure 4.7 shows optical absorption spectra calculated using Mie's theory for nanoparticles of different metals, with varying sizes. For most systems, the plasmon resonance remains relatively constant

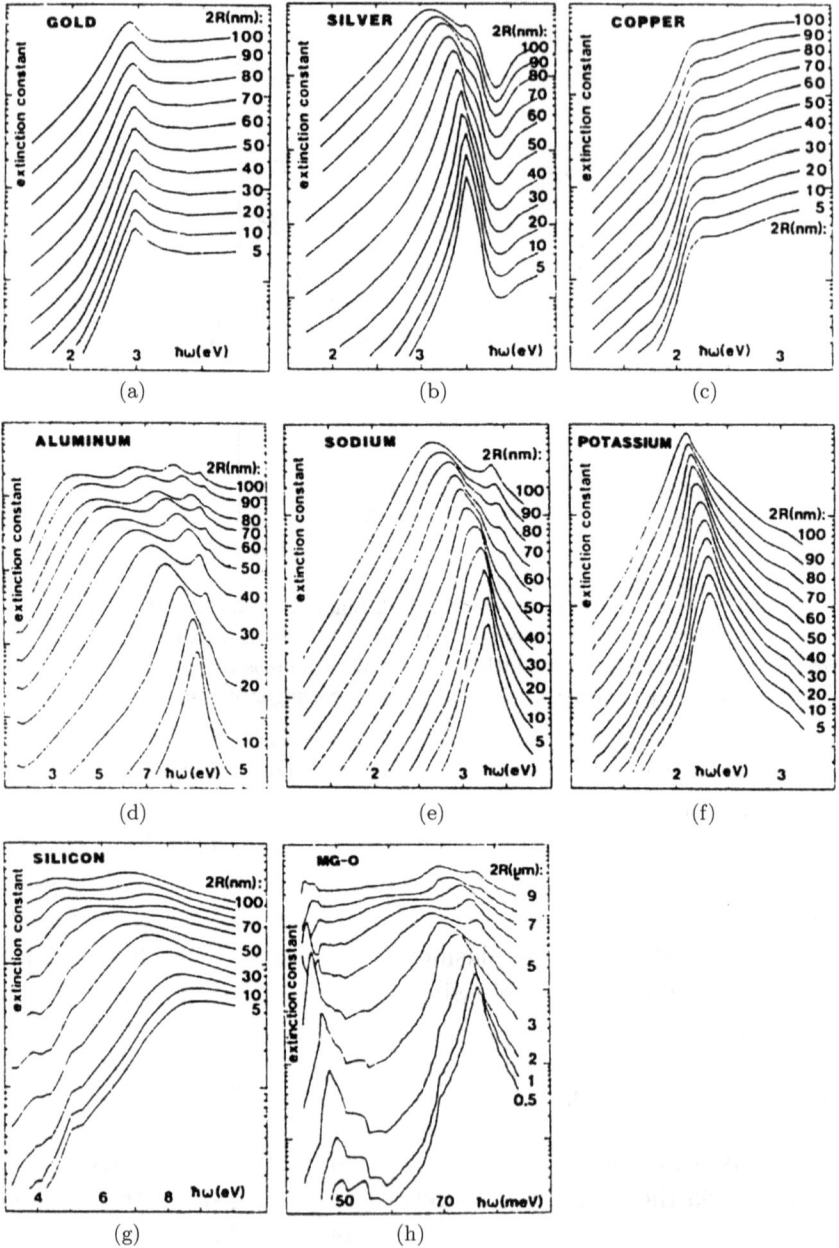

Figure 4.7:   Calculated absorption spectra of metallic nanoparticles of varying size and composition. Adapted with permission from [2]. © Springer Verlag.

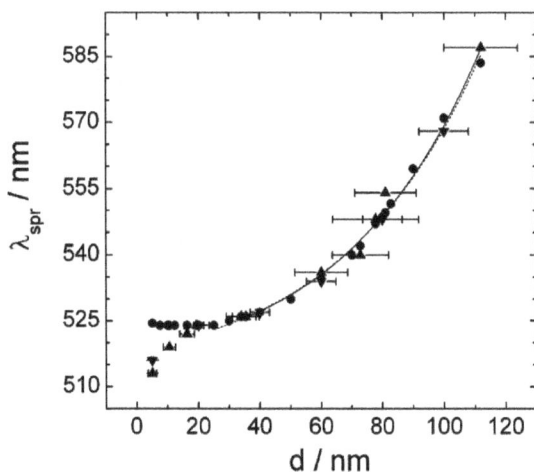

Figure 4.8:   Position of the absorption resonance maximum as a function of particle size for aqueous suspension of nanoparticles. Solid line: Fit to experimental data using Mie's theory. Adapted with permission from [3]. © American Chemical Society.

within the size range 5–30 nm, which is the most common for nanotechnology applications.

Recent data on aqueous suspensions of Au nanoparticles is reported in Figure 4.8 together with a fit based on Mie's scattering theory [3]. We note that size dependence remains very weak below about 40 nm. We also note a deviation from Mie's theory at very small sizes. This deviation might be due to surface defects, or interaction of surface atoms with surfactants. This interaction can alter the electron density N and with that, the position of the plasmon resonance.

4.5.3. *Shape dependence*

Another issue to bear in mind is that when particles are not spherical in shape the energy of the plasmon cannot be calculated analytically, and, most importantly, has a strong dependence on nanoparticle shape and size. A good way to explain this is to look at the changes in the absorption spectrum induced by nanoparticle coalescence. Figure 4.9 reports the results of a vacuum deposition experiment. Au

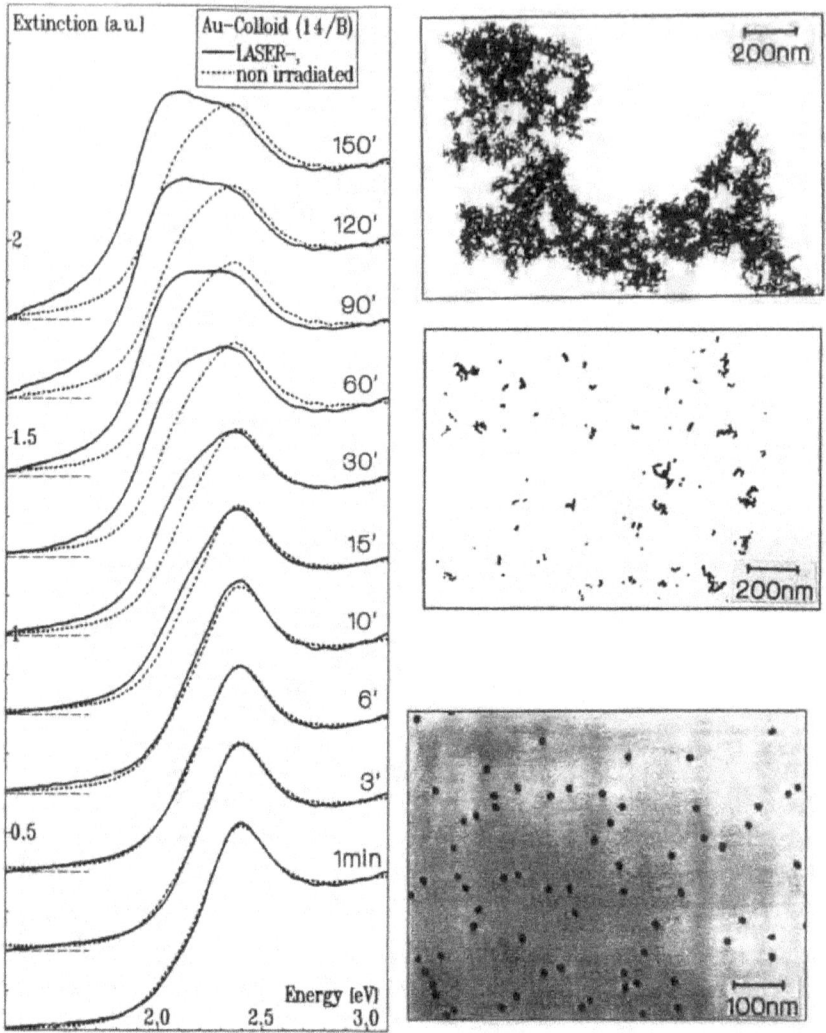

Figure 4.9: Illustration of the spectral shift induced by nanoparticle coalescence. Left: absorption spectra of films of Au nanoparticles. The films were deposited by vacuum evaporation for the times reported. Right: transmission electron microscopy of the films. Nanoparticles coalesce with incresing deposition time. Adapted with permission from [2]. © Springer Verlag.

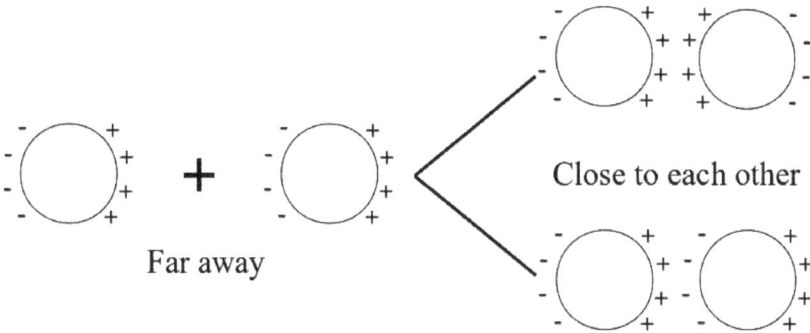

Figure 4.10: Schematic representation of lowering of the plasmon excitation energy by nanoparticles in close proximity.

nanoparticles were deposited in vacuum on a transparent substrate, and absorption spectra were measured for increasing deposition times. Transmission electron microscopy was used to characterize the film morphology. The absorption maximum shifts towards lower energies (i.e., longer wavelengths) with increasing deposition times. The reason for this is that coupling of dipoles lowers the energy of the system, as shown in Figure 4.10. The configuration to the lower right has less energy that the two isolated particles. Consequently, the energy necessary to excite a plasmon decreases when particles are close to each other.

More rigorously, one can calculate the interaction of dipole moments $p = eq$ in the same way as coupled harmonic oscillators. The Hamiltonian is: $H = \frac{1}{2} \sum_i constant \times [\frac{dq_i}{dt} + \omega_2^2 q_i^2] + e^2 \sum_{i,j} q_i T_{i,j}$. $q_j T_{j,i}$ is the dipole tensor: $T_{i,j} = \frac{\partial^2}{\partial^2 R_{ij}} \frac{1}{R_{ij}}$ and it has eigenvalues λ_μ. The table below reports the eigenvalues for different lengths of a linear chain. Note that the eigenvalue for longitudinal propagation decreases when the chain length increases. Correspondingly, the energy of the longitudinal plasmon oscillation decreases with increasing chain length.

For nanorods, we can identify two plasmonic modes. One, the transverse plasmon, is similar in energy to the surface plasmon of spherical nanoparticles. The longitudinal plasmon is at lower energies, and its energy depends on the length of the rod. A schematic

Table 4.1: Optically active modes of various linear cluster aggregates with their multiplicities. The second and third column give the optical absorption relative to the single sphere and the eigenvalues λ_μ, respectively. Adapted with permission from [2]. © Springer Verlag.

Pair		1×	2	−0.250
		2×	2	+0.125
Triplet		2×	0.065	−0.338
		2×	2.93	+0.185
		1×	2.93	−0.370
		1×	0.065	+0.338
Quadruplet		1×	3.83	−0.436
		1×	0.15	+0.176
		2×	0.16	−0.088
		2×	3.83	+0.218

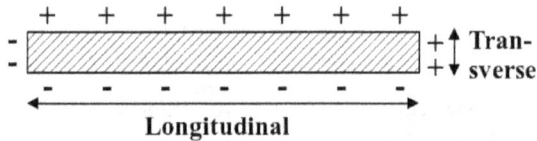

Figure 4.11: Schematic representation of the dipole arrangement in the transverse and longitudinal plasmons in a nanorod.

representation of the two plasmons is shown in Figure 4.11, and an example is presented in Figure 4.12. The absorption spectrum for spherical particles (a) presents an absorption maximum of around 400 nm. For nanorods, a second maximum is evident, which shifts

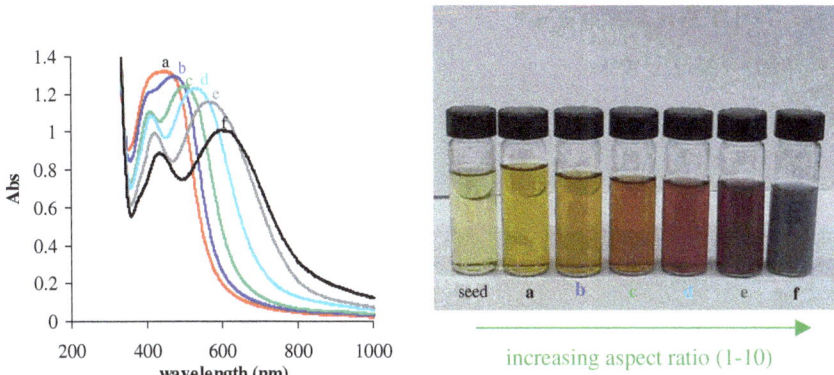

Figure 4.12: Left: absorption spectra of Ag nanowires with increasing aspect ratio. Right: digital camera images of the corresponding colloidal suspensions. Adapted with permission from [4]. © Wiley-VCH.

to lower energies with increasing aspect ratio. Note how the peak at ~400 nm persists throughout the series of spectra. This peak corresponds to the transverse plasmon (as well as residual spherical nanoparticles in the suspension) [4].

4.6. Optical properties of very small nanoparticles

Figure 4.8 shows that Mie's theory can be used to predict the position of the surface plasmon of spherical nanoparticles down to about 10 nm. Below this threshold, deviations appear. These deviations are due to the interaction between surfactants and surface, and an increased surface-to-volume ratio. The plasmon resonance depends on electron density (Eq. (4.27)), which is affected by the presence of surfactants and defects. At even smaller sizes ($\lesssim 2$ nm), nanoparticles exhibit completely different optical properties. For example, Fig. 4.13 shows the absorption spectra of Au_N clusters, $N = 8$–11. The spectra show absorption peaks at about 300, 400 and 500 nm. These peaks cannot be reconciled with Mie scattering and are due to the fact that clusters of this size behave, in essence, like molecules. *Ab initio* theoretical calculations show that HOMO-LUMO (that is, not plasmonic) transitions can account for the observed features, as shown in Figure 4.14.

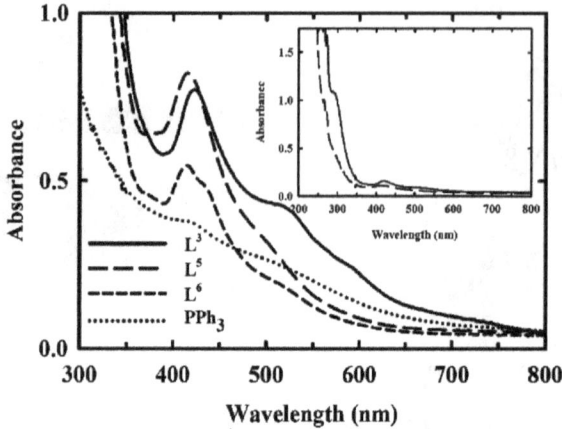

Figure 4.13: Optical absorption spectra of Au$_N$ clusters, $N = 8$–11. Adapted with permission from [5]. © American Chemical Society.

The transition from the small, molecular-like clusters to the plasmonic regime was investigated by the group of A. Dass. This group prepared thiolate-capped Au$_N$ clusters and separated their sizes using gel electrophoresis [7]. The absorption spectra of different chromatographic bands (i.e., different cluster size sizes) evolve from discrete to featureless, until a plasmonic peak presents itself for sizes around $N = 200$, as shown in Figure 4.15.

4.7. Surface plasmons

Let us go back to the dispersion relation in a material, Eq. (4.10): $\omega^2 = \frac{c^2}{\tilde{\epsilon}}k^2$. In vacuum, the dispersion relation is $\omega = ck$, which translates into $f = \frac{\omega}{\lambda}$ and tells us that the frequency decreases for increasing wavelengths. Now, Eq. (4.10) is subtly different from its vacuum counterpart. There, $\tilde{\epsilon}$ is IMAGINARY and also FREQUENCY-DEPENDENT, and this leads to interesting effects.

For starters, let us look at Eq. (4.23):

$$\epsilon_1(\omega) \approx 1 - \frac{\omega_p^2}{\omega^2} \to 1 \quad \text{when } \omega \gg \omega_p$$

$$\epsilon_2(\omega) \approx \frac{\omega_p\gamma}{\omega^3} \to 0 \quad \text{when } \omega \gg \omega_p.$$

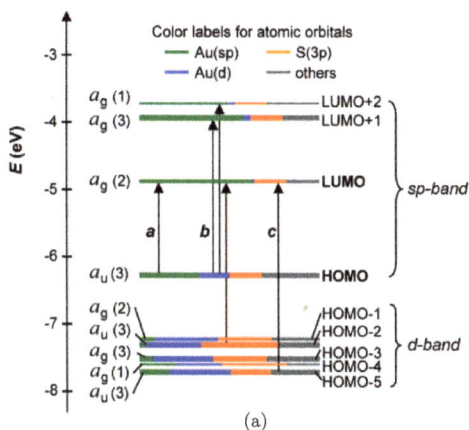

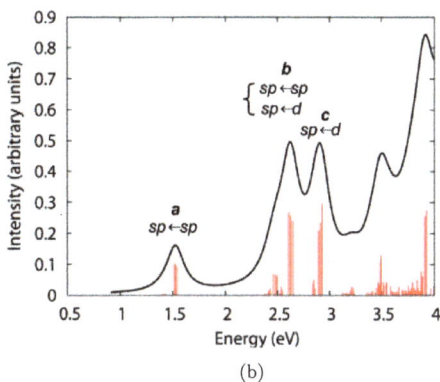

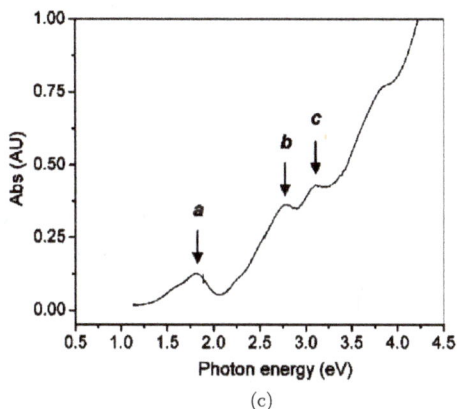

Figure 4.14: Theoretical simulations of the optical absorption of Au$_{25}$ clusters. (a) Orbitals and transitions. (b) Predicted transition intensities. (c) Experimental spectra. Adapted with permission from [6]. © American Chemical Society.

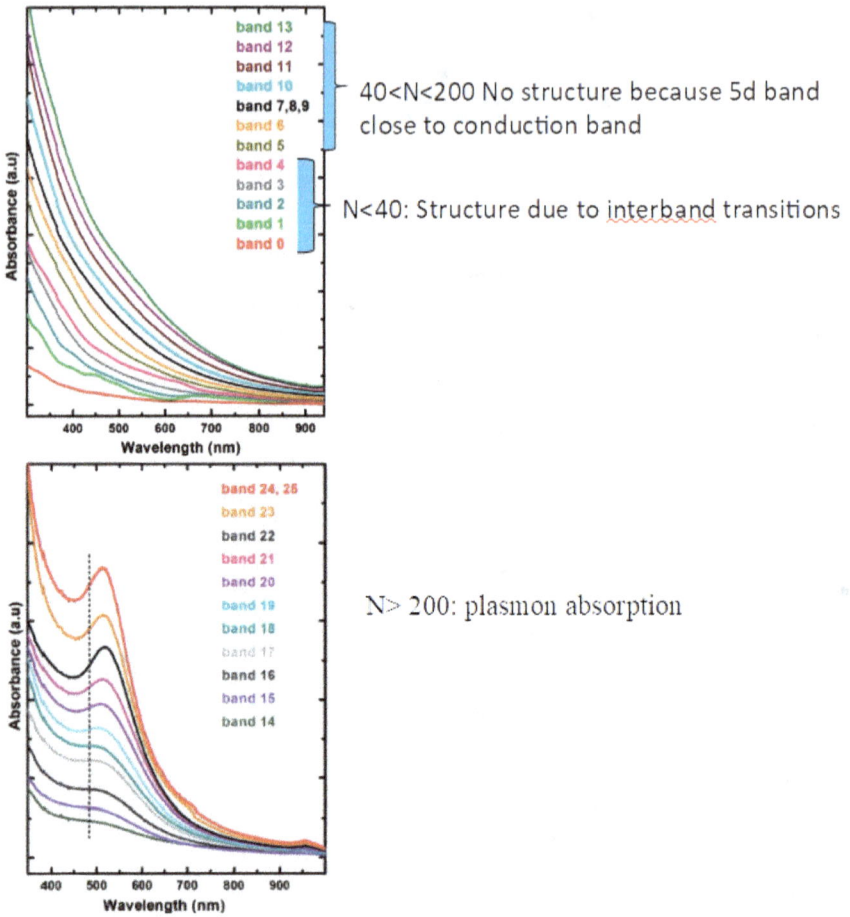

Figure 4.15: Evolution of optical absorption spectra as a function of cluster size. Adapted with permission from [7]. © Royal Society of Chemistry.

Let us work now for a system for which the imaginary part is negligible, say, Ag in the blue region of the visible spectrum, and let us substitute Eq. (4.23) into Eq. (4.10). We then obtain:

$$\omega^2 = \omega_p^2 + c^2 k^2, \tag{4.29}$$

which yields the curve in Figure 4.16.

In Figure 4.16, the forbidden frequency range indicates frequencies for which the wave cannot propagate through the bulk; see also the

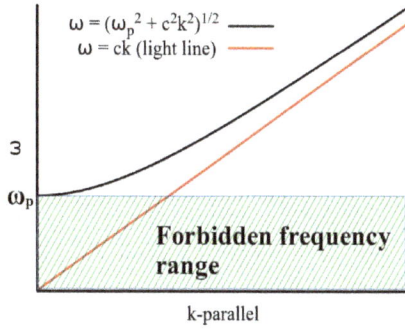

Figure 4.16: Plot of the dispersion relation, Eq. (4.29).

discussion of Eq. (4.23). The straight line is the dispersion relation in vacuum, which is also called the "light line".

Now, this type of system is not particularly interesting, since it is rare to work at frequencies larger than ω_p. Yet, the example is useful to introduce the dispersion relation. Now comes the crux of the problem, which we use to make sensors. It turns out that in the forbidden region waves will not propagate through the bulk of a solid, but they can (and do) propagate as evanescent waves on the surface.

The geometry of the problem is shown in Figure 4.17. That is, an incident wave can be written as $E = Ae^{i(k_x x + k_z z - \omega t)}$. Using standard electromagnetism theory, one can derive a dispersion relation for k_x:

$$k_x = \frac{\omega}{c}\left[\frac{\left(1 - \frac{\omega_p^2}{\omega^2}\epsilon_2\right)}{1 - \frac{\omega_p^2}{\omega^2} + \epsilon_2}\right]^{1/2}, \tag{4.30}$$

which is plotted in Figure 4.18 (black curve).

We note that the dispersion relation for k_x has an asymptotic value $(k_x \to \infty)$ which is reached when the denominator of Eq. (4.30) goes to zero:

$$1 - \frac{\omega_p^2}{\omega^2} + \epsilon_2 = 0, \quad \text{that is } \omega = \frac{\omega_p}{\sqrt{1 + \epsilon_2}} \tag{4.31}$$

(Note the similarity, but also the difference, with the resonance condition for the nanoparticle plasmon). Physically, the surface wave looks as in Figure 4.19.

Introduction to Nanotechnology

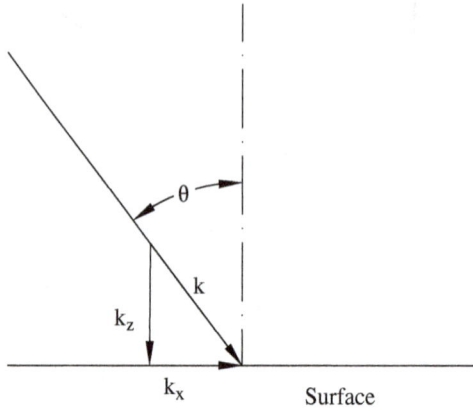

Figure 4.17: Incident wave vector (k), perpendicular wave vector (k_z) and parallel wave vector (k_x) for an electromagnetic wave incident at an angle θ onto a surface.

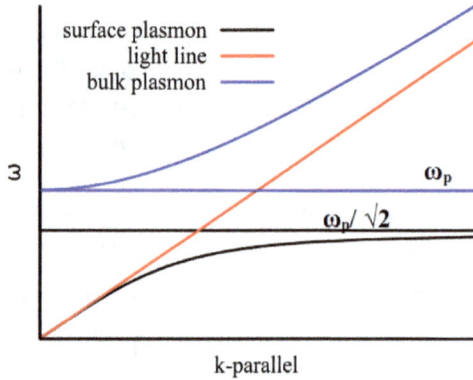

Figure 4.18: Bulk and surface plasmon dispersion curves.

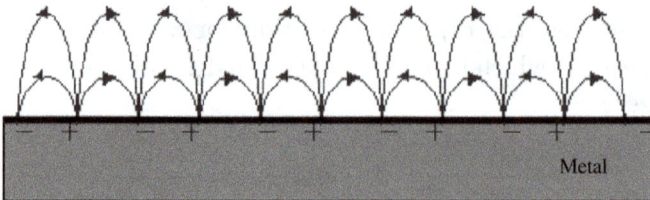

Figure 4.19: Representation of surface dipoles for an evanescent surface wave. Adapted with permission from Pitarke *et al.* (2006); © IOP Publishing.

4.7.1. *Surface plasmon sensors*

The surface plasmon concept is extremely powerful for sensor applications. Before we go into sensors, we have to notice that usually photons have a small k. If you look at the dispersion relation in Figure 4.18 you will see that small k means small ω and this is a headache, since, ideally, we want to work with visible light. However, from the geometry of the problem shown in Figure 4.17, we notice that $k_x = k\sin(\theta)$. To increase k_x, one needs to work at a grazing angle. This can be attained by using a total internal reflection configuration. That is, one uses the evanescent wave on the surface of a prism operated under total internal reflection conditions to couple to the evanescent surface plasmon wave. The prism can be put in close proximity of a metal film, or a metal film can be evaporated on it.

The working principle of a surface plasmon sensor is as follows. The beam is incident with a fixed frequency ω. The incidence angle is varied to span a range of values of k_x, which corresponds to the red line in Figure 4.20. The plasmonic film is in contact with a matrix with dielectric constant ϵ_m. Let us assume that ϵ_m has initially a low value (say, air). The dispersion relation of the surface plasmon is given by the black line and has an asymptotic value given by Eq. (4.31). At angle A the curve of the incident light intersects the dispersion curve of the surface plasmon. The incident light couples with the plasmon and the detector measures a minimum in the intensity of the reflected light. When the dielectric constant of the matrix is changed, the resonance angle shifts to B. Since angle changes are easily measured, surface plasmon sensors are very sensitive and convenient to use. An example is shown in Figure 4.21.

4.8. Beating the diffraction limit using plasmonics

There is another way of adding parallel momentum to the incident light: use of a diffraction grating, as shown in Figure 4.22. Diffraction adds a parallel momentum of $\frac{2n\pi}{a}$, where a is the period of the grating.

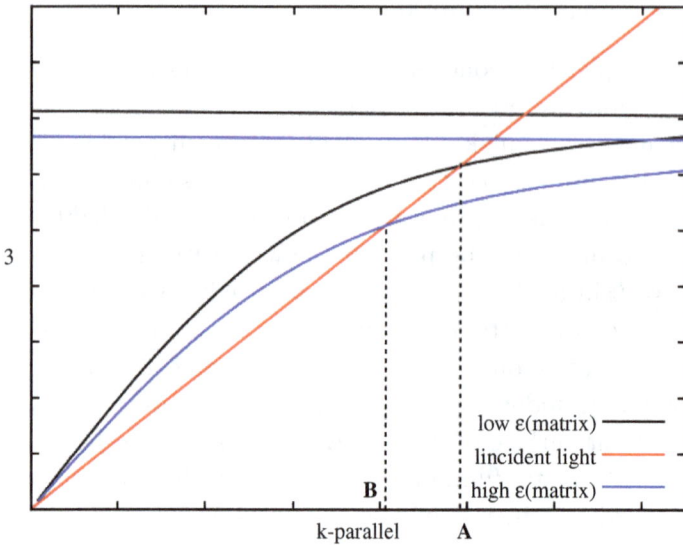

Figure 4.20: Schematic representation of the working principle of a surface plasmon sensor. The red line indicates the dispersion of the incident wave. The resonance angle (k-parallel) changes from position A to position B when the index of refraction of the matrix is increased.

Thus, the surface plasmon will be excited when:

$$k_{sp} = k_x + \frac{2n\pi}{a}$$

$$\omega_{sp} = \omega_i, \tag{4.32}$$

where ω_i is the frequency of the incident light.

One could measure the coupling of light to the plasmon by placing a grating on top of a film, as shown in Figure 4.23. While such a set-up would work, it would be impractical, since detection would be at a grazing angle. A more practical configuration is to etch a grating into the film, as shown in Figure 4.24. Plasmon coupling is detected by illuminating the grating with white light at normal incidence and measuring the light transmitted through the grating. The results for a grating with holes 150 nm wide, 900 nm apart are shown in Figure 4.25. Minima correspond to coupling of the incident light to

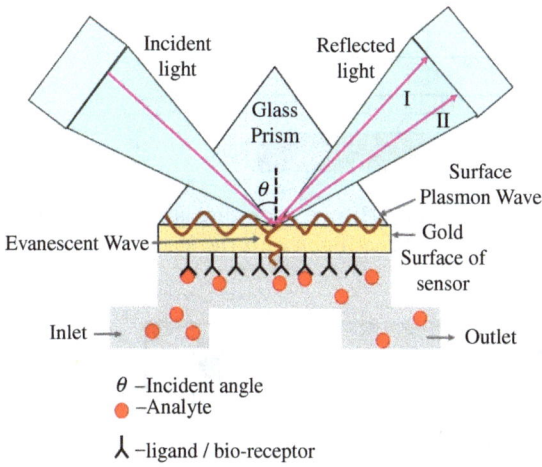 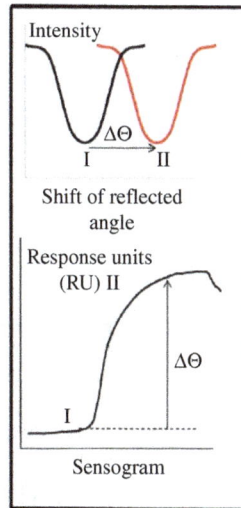

(a) (b)

Figure 4.21: Application of a surface plasmon sensor to a biological system. The plasmonic film is derivatized with a target substance, and the resonance is as indicated by the red curve. When an antibody capable of attaching to the target is introduced into the system, the index of refraction of the matrix changes, and so does the resonance angle. Adapted with permission from Mehrotra *et al.* (2019); © MDPI.

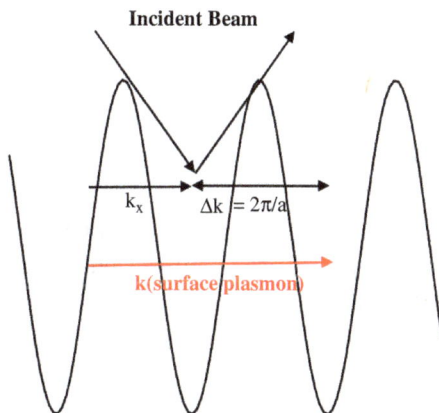

Figure 4.22: Use of a diffraction grating to increase the parallel momentum of incident light.

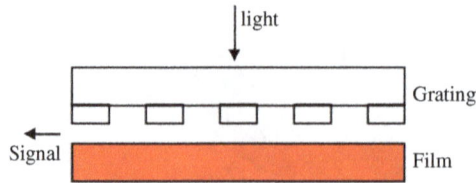

Figure 4.23:   A possible plasmon coupling scheme. A grating is placed on top of an Ag film to monitor coupling of the incident light with the surface plasmon.

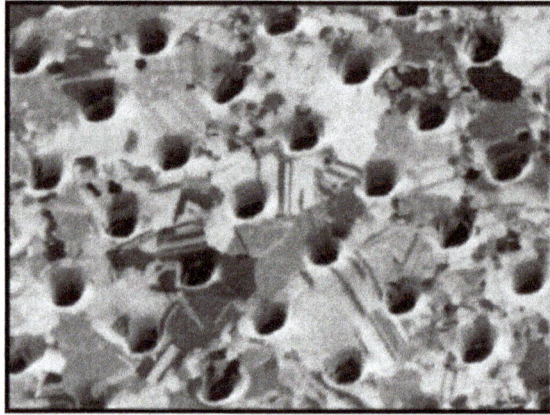

Figure 4.24:   Scanning electron microscope image of a grating creatd within a Ag film by focused ion beam milling. Adapted with permission from [8]. © American Physical Society.

the surface plasmon. Their position is calculated using Eq. (4.32) and is indicated by solid lines in the figure.

Surface plasmon coupling can be used to beat the diffraction limit as follows [9]. A metal grating is prepared as a part of a lithographic mask, and placed in contact with a photoresist, as shown in Figure 4.26. The mask is then illuminated with ultraviolet light (365 nm), which polymerizes the photoresist. The grating is designed so that the diffracted light couples with the surface plasmon. That way, light is transmitted through the holes. Diffracted light, which would broaden the patterns, is rejected.

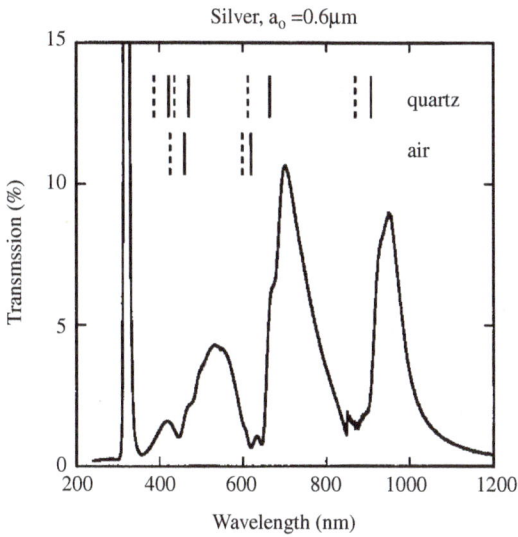

Figure 4.25: Transmission of white light through the grating of Figure 4.24. Minima correspond to coupling of the incident light to the surface plasmon. Adapted with permission from [8]. © American Physical Society.

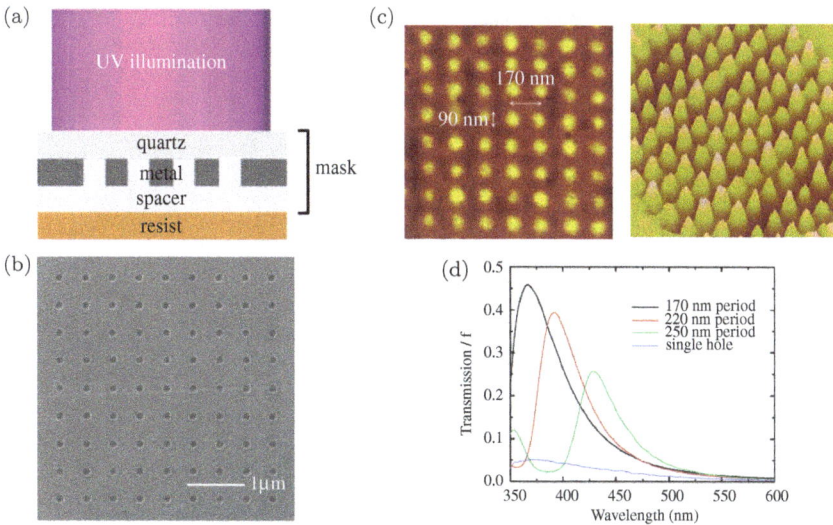

Figure 4.26: (a) Experimental configuration for plasmonic grating lithography. (b) Grating image. (c) Top view of the exposed and developed photoresist and the surface profile of patterned substrate. (d) Transmission through the grating for the indicated wavelengths and grating geometry. Adapted with permission from [9]. © American Chemical Society.

References

1. See, for example, "Light scattering by small particles". By H. C. van de Hulst. New York (John Wiley and Sons), London (Chapman and Hall), 1957.
2. U. Kreibig, M. Vollmer, "Optical Properties of Metal Clusters". Berlin: Springer-Verlag (1995).
3. W. Haiss, N. T. K. Thanh, J. Aveyard, D. G. Fernig, *Anal. Chem.* **79**, 4215–4221 (2007).
4. C. J. Murphy, N. R. Jana, *Adv. Mater.* **14**, 80–82 (2002).
5. M. F. Bertino, Z. M. Sun, R. Zhang, and L. S. Wang, *J. Phys. Chem. B.* **110**, 21416–21418 (2006).
6. M. Zhu, C. M. Aikens, F. J. Hollander, G. C. Schatz, R. Jin, *J. Am. Chem. Soc.*, **130**, 18, 5883–5885 (2008).
7. N. Kothalawala, J. L. West IV, and A. Dass, *Nanoscale*, **6**, 683 (2014).
8. H. F. Ghaemi, Tineke Thio, and D. E. Grupp T. W. Ebbesen H. J. Lezec, *Phys. Rev. B*, **58**, 6779 (1998).
9. W. Srituravanich, N. Fang, C. Sun, Q. Luo, and X. Zhang, *Nano Lett.*, **4**, 6 (2004).

Chapter 5

Magnetism

Magnetism is a quite complicated phenomenon which can result in several different spin arrangements. Some of the most common are listed below and shown in Figure 5.1:

- Paramagnetism
- Ferromagnetism
- Ferrimagnetism
- Antiferromagnetism

5.1. A bit of theory

To get familiar with magnetic effects, let us ask a question. Fe, Ni and Co are magnetic. However, the elements in the rows below them in the periodic table (Pt, Pd, etc.) are not. The reason is the exchange interaction. Its theoretical treatment resembles that of the He atom and of the hydrogen molecule. That is, we have electrons that may or may not be on the same orbital (or on the same atom) and may or may not have the same spin. We want to calculate the energy of the different configurations.

Let us start from a system like H_2^+, that is, two attractors, one left (L), one right (R) and one electron in between, as shown in Figure 5.2.

The Schrödinger equation for this system is:

$$E\psi = \frac{-\hbar^2}{2m}\nabla^2\psi + V_0(r)\psi + V_0(|\mathbf{r} - \mathbf{R}|)\psi, \qquad (5.1)$$

Paramagnetic

Ferromagnetic

Ferrimagnetic

Antiferromagnetic

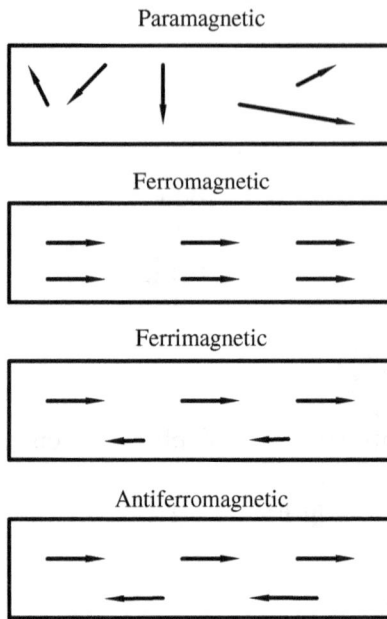

Figure 5.1: Common types of magnetism in solids, and the corresponding spin arrangements.

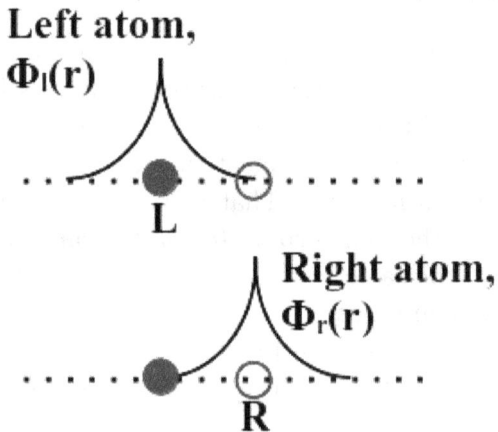

Left atom,
$\Phi_l(r)$

L

Right atom,
$\Phi_r(r)$

R

Figure 5.2: Atomic wave functions of the diatomic pair model.

where $V_0(r) \approx \frac{Z}{r}$ is the atomic potential of the first atom and $V_0(|r - R|)$ is the potential from the second atom. The solution is a linear combination of one wave function centered on the left and one on the right atom: $\phi_l(r) = \phi_0(r), \phi_r(r) = \phi_0(|r - R|)$. For ϕ_0 one can use, for example, atomic-like wave functions: $\phi_0 \approx e^{-(\frac{r}{R_0})}$. Under these assumptions, the Hamiltonian has the form:

$$\begin{pmatrix} E_0 & t \\ t & E_0 \end{pmatrix},$$

(5.2)

where E_0 is the energy of the molecule and t is the hopping integral:

$$t = \int \phi_0^*(r) V_0(|\mathbf{r} - \mathbf{R}|) \phi_0(|\mathbf{r} - \mathbf{R}|) dV.$$

(5.3)

This integral plays a vital part in the discussion of magnetism. It represents, approximately, the probability that the electron hops from one atom to the other. In other approximations, the integral has an even more intuitive form, and it represents the overlap of wavefunctions located on different atoms.

$$t = \int \phi_l(\mathbf{r}) \phi_r(\mathbf{r}) dV.$$

(5.4)

Diagonalization of the Hamiltonian (5.2) leads to two solutions, one symmetric and the other antisymmetric:

$$\psi_s = \phi_l + \phi_r \quad E = E_0 + t$$
$$\psi_a = \phi_l - \phi_r \quad E = E_0 - t,$$

(5.5)

which are shown in Figure 5.3.

For a two-electron system one has to consider spins and also the Coulomb interaction between electrons:

$$V_c = \frac{e^2}{|\mathbf{r} - \mathbf{r}'|},$$

(5.6)

Antisymmetric, $\psi_a(r)$

Symmetric, $\psi_s(r)$

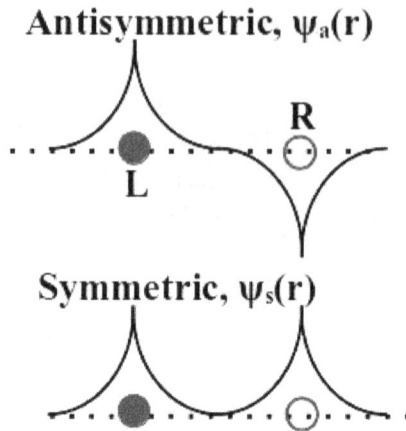

Figure 5.3:   Symmetric and antisymmetric wavefunctions used in Eq. (5.5).

where $\mathbf{r}$ is the position of the first electron and $\mathbf{r}'$ is the position of the second electron. The Hamiltonian is

$$H = H_0(\mathbf{r}) + H_0(\mathbf{r}') + V_c, \tag{5.7}$$

where

$$H_0 = \frac{-\hbar^2}{2m}\nabla^2 + V_0(\mathbf{r}) + V_0(|\mathbf{r} - \mathbf{R}|). \tag{5.8}$$

The wavefunction will be a linear combination of the atomic wavefunctions centered on each of the atoms:

$$\psi = c_1\phi_L(r)\phi_R(r') + c_2\phi_L(r)\phi_R(r')$$
$$+ c_3\phi_R(r)\phi_L(r') + c_4\phi_R(r)\phi_R(r'), \tag{5.9}$$

and the solutions are obtained by diagonalization of the matrix:

$$H = 2E_0 \begin{pmatrix} U & t & t & J_D \\ t & 0 & J_D & t \\ t & J_D & 0 & t \\ J_D & t & t & U \end{pmatrix}, \tag{5.10}$$

where

$$U = \iint \phi_L^*(\mathbf{r})\phi_L^*(\mathbf{r}')V_c(\mathbf{r},\mathbf{r}')\phi_L(\mathbf{r})\phi_L(\mathbf{r}')$$

$$J_D = \iint \phi_L^*(\mathbf{r})\phi_R^*(\mathbf{r}')V_c(\mathbf{r},\mathbf{r}')\phi_R(\mathbf{r})\phi_L(\mathbf{r}'). \tag{5.11}$$

U is the Coulomb integral, and J_D is called direct exchange. The energies of the Hamiltonian are:

$$E_1 = 2E_0 - J_D$$

$$E_2 = 2E_0 + U - J_D$$

$$E_3 = 2E_0 + U/2 + J_D - \sqrt{4t^2 + \frac{U^2}{4}} \cdot \tag{5.12}$$

$$E_4 = 2E_0 + U/2 + J_D + \sqrt{4t^2 + \frac{U^2}{4}}.$$

Let us now discuss the relevance of the energy terms.

Usually: $J_D \ll U \approx t$. t depends on the diameter of an orbital but ALSO on the interatomic distance. In fact, in Eq (5.4) V is an attractive potential that decreases at large values of $|r - R|$. As for the wavefunctions, the most important for the discussion are those with the lowest energies:

$$|1\rangle = \frac{1}{\sqrt{2}}|LR\rangle - \frac{1}{\sqrt{2}}|RL\rangle$$

$$|3\rangle = \frac{\sin\chi}{\sqrt{2}}(|LL\rangle + |RR\rangle) + \frac{\cos\chi}{\sqrt{2}}(|LR\rangle + |RL\rangle), \tag{5.13}$$

where
$$|LL\rangle = \phi_L(r)\phi_L(r'), \text{etc.}$$

$$\tan(2\chi) = \frac{-4t}{U}$$

Of these wavefunctions, $|1\rangle$ is antisymmetric. Therefore its spin part is symmetric: $|\uparrow\uparrow\rangle$ = ferromagnetic. Wavefunction $|3\rangle$ is symmetric; therefore its spin part is antisymmetric and therefore antiferromagnetic.

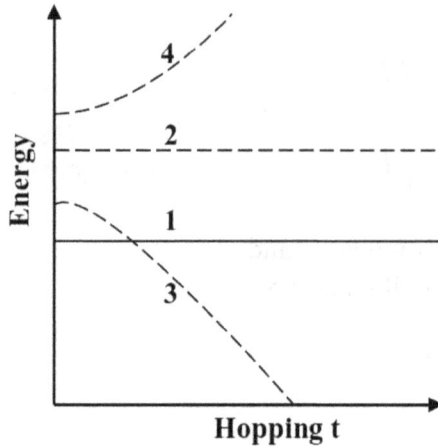

Figure 5.4: Dependence of the energy of the eigenstates (Eq. (5.12)) on the hopping integral.

Figure 5.4 shows the dependence of the energies of the ferro- and antiferro-magnetic components as a function of the overlap integral t.

We see that the antiferromagnetic state, $|1\rangle$, is favored at small t. The physical reason is that large t means that both electrons will tend to be on the same atom, and this will force the spins into an anti-symmetric (antiparallel) state. This argument explains why the light transition metals (Co, Fe, Ni) are magnetic, while heavier metals (Pt, Pd, etc.) are not. For all these metals (light and heavy) the lattice spacing is pretty much the same. However, the radius of the valence (d) electrons increases in the heavier metals. Therefore, the overlap integral increases, and magnetism decreases, when the row in the periodic table increases.

5.2. Domain formation

In nanotechnology, domain formation plays a paramount role. To understand domain formation, we have to recall some very fundamental concepts of electromagnetism. Maxwell's equation relating the magnetic field and a current, in its integral form, reads:

$$\oint \mathbf{B} \cdot dl = \mu_0 \int \mathbf{J} \cdot d\mathbf{A} = \mu_o I. \qquad (5.14)$$

Let us now calculate the magnetic field surrounding a conductor carrying a current I, as shown in Figure 5.5. At a distance R from the conductor:

$$\oint \mathbf{B} \cdot dl = 2\pi R B = \mu_0 I, \tag{5.15}$$

that is,

$$B = \frac{\mu_0 I}{2\pi R}. \tag{5.16}$$

We now introduce the H field:

$$H = \frac{B}{\mu_0} = \frac{I}{2\pi R}. \tag{5.17}$$

The field H is a useful quantity, since it represents the field generated outside a magnet, for example by an external current. Let us now consider the case of a solenoid with N windings, each of which with a length l. Some additional math shows that the magnetic field is:

$$B = \mu_0 H$$
$$H = \frac{NI}{l}. \tag{5.18}$$

Now let us come closer to nanotechnology. Permanent magnets can be approximated by a bunch of current loops, each the size of an

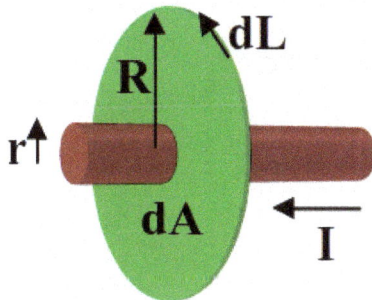

Figure 5.5: Geometry used for the calculation of the magnetic field surrounding a wire carrying a current I.

atom. Before looking at the magnetization of the material, we write Eq. (5.18) in a slightly different way:

$$B = \frac{\mu_0 N I}{l}$$

$$B = \mu_0(H + M),$$ (5.19)

where the second part of Eq. (5.19) reminds us that the field B is the sum of the external field H and of the magnetization of the material, M. Assuming $H = 0$ (no external fields), rearrangement of Eq. (5.19) yields:

$$B = \mu_0 I A n,$$ (5.20)

where A is the area of an atomic loop and n the number of loops per unit volume. In other words, in Eq. (5.19), we assume to switch off the field due to external currents (H). What remains is the field (M) due to internal currents (atomic loops). For a hydrogen atom:

$$I A = \mu_m = e\frac{\omega}{2\pi}\pi r_0^2,$$ (5.21)

where ω is the angular velocity of the electron around the nucleus, r_0 the Bohr radius and μ_m indicates the magnetic moment of a single atom. Using tabulated values we obtain: $I A = 9.27 \times 10^{-24}$ A $\cdot$ m^2. Typical atomic densities in metals are $n = 10^{29}$ m^{-3}; hence the magnetization of an (idealized) assembly of hydrogen atoms is $B \sim 1$ T. While crude, this approximation compares well with the saturation magnetization of early transition metals:

Metal	Saturation magnetization (T)
Fe	2.2
Co	1.7
Ni	0.6

Let us now consider a slab of a uniformly magnetized material, as shown in Figure 5.6. The magnet can be envisaged as a series of

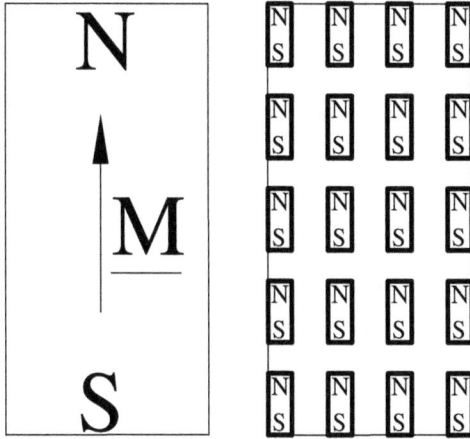

Figure 5.6: Idealized representation of a magnetized system, and its breakdown into aligned dipoles.

aligned dipoles. The surface dipoles cause trouble. Namely, the South pole of the dipoles at the bottom does not talk to another dipole, and neither does the North pole of the top row of dipoles. These "un-compensated" dipoles behave like electric surface charges and generate a field inside the magnet that has the *opposite* direction as the magnetization of the material, as shown in Figure 5.7. This field (H_d) is a *demagnetizing* field and plays an incredibly important role in nanostructured magnetic materials.

The demagnetization field H_d generated by the surface dipoles induces an energetic penalty, which can be calculated as follows. From the general theory of electromagnetism, we know that the energy U of a magnetic dipole μ_m in a field $B = \mu_0 H$ is $U = -\mu_m.B$. If the system has n dipoles per unit volume, the magnetization M is defined as

$$M = \frac{\text{magnetic moment}}{\text{unit volume}} = \mu_m \cdot n, \qquad (5.22)$$

And the interaction energy (per unit volume) is:

$$u = \mathbf{M} \cdot \mathbf{B}. \qquad (5.23)$$

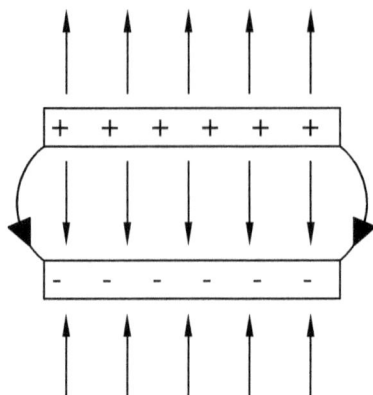

Figure 5.7: Field generated by the surface dipole charges.

Starting from Eq. (5.23), basic electromagnetic theory (see, for example Ref. [1]) can be used to yield a quite interesting result:

$$u = \frac{\mu_o F M^2}{2},$$ (5.24)

where F is a factor, called demagnetizing factor, which depends on the shape of the particle. Analytical models for F can be found in Ref. [1]. An interesting finding of these calculations is that F is minimum for elongated ellipsoids, which explains why these shapes are preferred for certain magnet applications.

Let us now wrap up our discussion up and explain why domains form in magnetic systems. Let us assume that the sample has zero magnetization. Locally, the moments will align, because alignment lowers their energy. Since the magnetization must be zero, different regions of the sample will have magnetization pointing in different directions, so that the average is zero. Figure 5.8 shows two possible arrangements. In the one on the left, the magnet is divided into two regions with magnetization in opposite directions. This configuration yields two layers of surface charges. Let us now look at the right panel of Figure 5.8. The magnetization is also zero, but now the surface charges from one domain are next to the charges from another domain, and the charges are opposite in sign. This

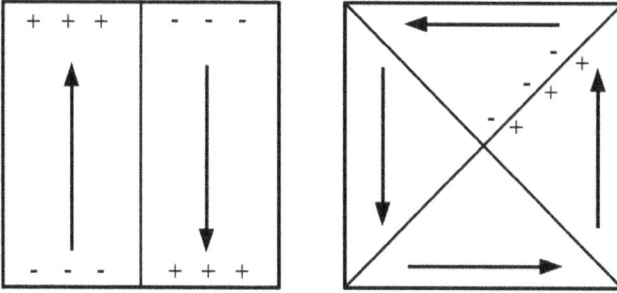

Figure 5.8: Domain formation in a sample with zero magnetization. Left: configuration with high magnetostatic energy. Right: configuration with lower magnetostatic energy.

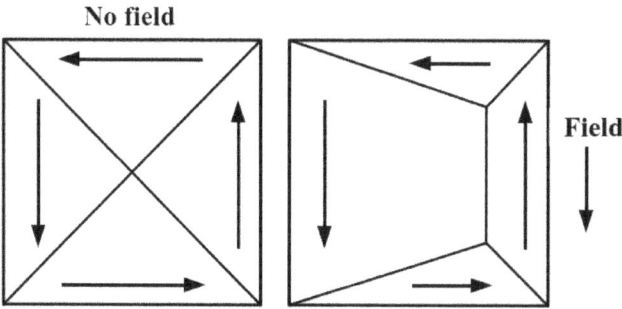

Figure 5.9: Domain growth upon application of an external field.

configuration reduces the amount of free surface charges and the energy penalty.

5.3. Going nano-

When an external field is applied to a magnetic system, domains which have a favorable orientation grow at the expense of the others, as shown in Figure 5.9. The magnetization of the sample therefore grows with increasing applied field, until it saturates. The saturation magnetization is denoted as M_s. When the direction of the field is reversed, the magnetization decreases (domain growth is reversed). However, when the applied field is zero, the magnetization

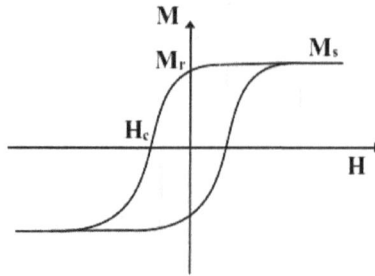

Figure 5.10:   Example of hysteresis curve.

is non-zero, as shown in Figure 5.10. This non-zero value is due to the fact that magnetic ordering is energetically advantageous, that is, a system wants to remain magnetized. The magnetization at zero applied field is called remanent magnetization and is indicated by M_r. When the field direction is reversed, magnetization keeps decreasing until it reaches zero. The field necessary to demagnetize the system is called coercive field and it is indicated by H_c.

5.3.1. *Single domain nanoparticles*

Let us recall the conclusion of Section 5.2: domains are formed in magnetic systems. Alas, boundaries between domains, being surfaces, cost energy. If a particle is small, it already will have to pay a hefty price to surface energy. If domains were also present, the energetic price would be prohibitive. This is why particles below a certain critical size tend to be single-domain. These single-domain (SD) particles are technologically very relevant, since they have high coercivities. The reason for the high coercivity is as follows. In multi-domain particles, magnetization can be increased (or decreased) by moving walls around, or by removing walls. Both processes cost energy. It turns out that creating domain boundaries is energetically expensive, while moving boundaries across a system is comparatively inexpensive. In SD particles, magnetization can only be changed by flipping all the spins at once, which is a VERY expensive business. One of the main applications of single-domain particles is for memory applications.

Magnetic recording is currently being replaced by flash memories; hence single-domain particles have lost a bit of their appeal. Due to their high coercivity, though, they remain important for permanent magnet applications. An in-depth review of the physical principles of single-domain particles is reported in Ref. [2].

5.3.2. *Superparamagnetism*

When the particle size is reduced further below the single domain size, another threshold is reached where coercivity goes to zero. This is the superparamagnetic limit. This happens because in superparamagnetic particles the number of spins is so small that thermal energy is sufficient to flip the magnetization.

To quantify superparamagnetism, we need to look first at anisotropy. Assume that you have a solid with a certain direction of symmetry, and assume that all spins point along that direction:

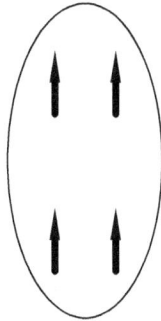

Now assume that for some reason the spins are rotated by an angle θ with respect to the "natural" alignment axis:

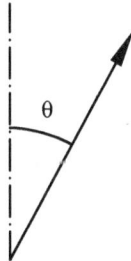

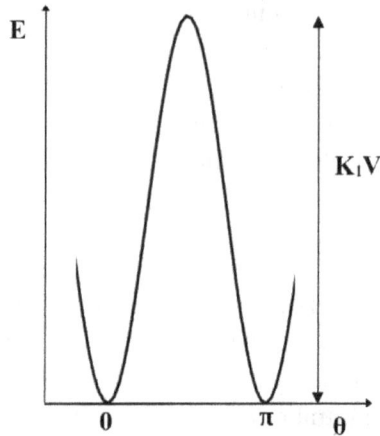

Figure 5.11: Relationship between the energy of the system and the angle between magnetization and minimum energy orientation.

Assuming this configuration will cost energy:

$$E_a = K_1 V \sin^2\theta, \tag{5.25}$$

where K_1 is a constant, V is the volume of the particle and $\sin^2\theta$ expresses the fact that 0 and π are equivalent angles, energetically speaking, as shown in Figure 5.11.

$K_1 V$ expresses the energy required to flip the spins between the two energetically equivalent positions of 0 and π. We notice that this energy depends on the volume of the particle. Therefore, when

$$K_1 V \approx k_B T, \tag{5.26}$$

the magnetization can flip spontaneously from one direction to its opposite. Hence, the net magnetization of a system of such particles is zero. The situation changes when a magnetic field is applied, say, along the $\theta = 0$ direction. In that case, particles with the spin oriented in the $\theta = \pi$ direction pay an energy price, as shown in Figure 5.12. Since the two directions are not equivalent, particles with the spin in the "wrong" direction will tend to flip. Therefore, the magnetization of a superparamagnet is not zero when a field is applied, and it increases with the applied field. The typical magnetization

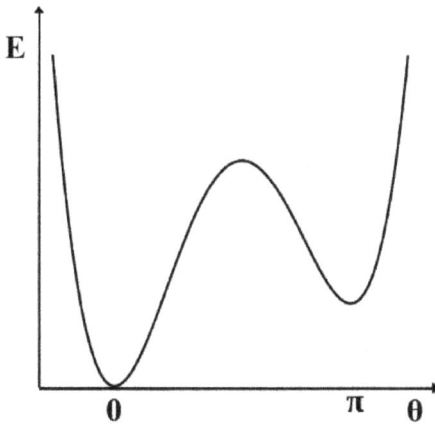

Figure 5.12: Introduction of a magnetic field breaks the symmetry between the $\theta = 0$ and $\theta = \pi$ spin orientiations.

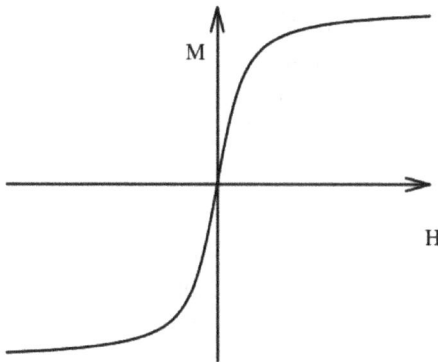

Figure 5.13: Typical M-H curve of a superparamagnet.

curve of a superparamagnet is reported in Figure 5.13. Superparamagnetic particles may appear "useless". Yet, they are needed for high frequency applications. For these applications, multi-domain nanoparticles are not desirable. Each time the field direction changes, domain walls have to be moved, resulting in heat generation. If single-domain, but not superparamagnetic, nanoparticles are employed, a large number of spins must be flipped each time the field direction changes. Flipping a large number of spins requires a large energy, reducing the attractiveness of this type of particles. Superparamagnetic nanoparticles represent therefore the more promising alley for high frequency applications, typically as microwave absorbers. They are also employed in biomedical applications such as hyperthermia and MRI contrast agents.

5.3.3. *Size dependence in superparamagnetism*

A powerful characterization method of superparamagnetic nanoparticles is the measurement of the dependence of magnetization on temperature. For this, two types of measurements are commonly employed. In one measurement, the particles are cooled to a low temperature ($\sim$50 K for most instruments) without an applied external field (Zero-Field-Cooling, ZFC). A field is then applied, the temperature is raised and the magnetization measured at each temperature, as shown in Figure 5.14 [3, 4].

In the second measurement scheme, the sample is cooled but this time a field is applied (Field-Cooling, FC). The temperature is then raised and the magnetization measured. To understand what happens when the temperature is raised, one has to refer to Figure 5.12. In the ZFC measurements, the two potential wells are equivalent during cooling (energetically speaking). The magnetization is zero at all temperatures. When a field is applied one of the wells becomes energetically unfavorable. At low temperatures, however, the spins do not have enough thermal energy to overcome the energetic barrier between the two wells. Hence, the magnetization does not change and remains zero. When the temperature is raised, the transition probability increases, particles "jump" to the lower potential well and the magnetization becomes nonzero. At very high temperatures the magnetization decreases, since there is now enough thermal energy for particles to jump back to the unfavorable configuration. The overlap of magnetization-increasing and decreasing- contributions yields the maximum observed in the ZFC curves of Figure 5.14. As for the FC curves, the magnetization is high at low temperatures, since in the

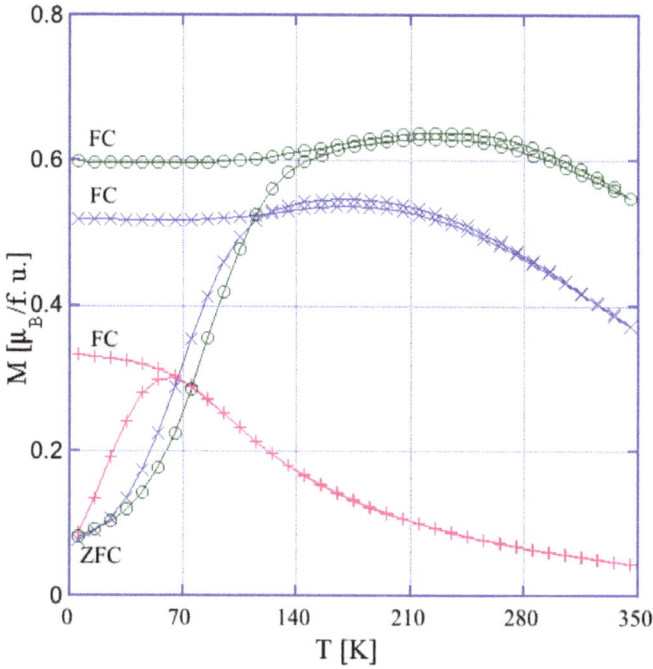

Figure 5.14: ZFC and FC curves for three γ-Fe$_2$O$_3$ nanocrystalline samples, 6.1 nm ($+$), 11.8 nm ($\times$), and 15.0 nm ($\circ$). Adapted with permission from [3]. © IOP Publishing.

initial stages of cooling there is enough energy for the spins to flip to the favorable orientation. The magnetization decreases when the temperature is increased, since the spins have now enough energy to jump to the unfavorable configuration. Comparison of FC and ZFC curves provides information on the energetics of the system. One should also note that the maximum magnetization and the maximum magnetization temperature of ZFC curves depend on particle size, as shown in Figure 5.15. The interpretation of the size dependence is as follows. The temperature for which the magnetization is highest increases with increasing particle size. This happens for a simple reason: the barrier is proportional to the volume of the particle, Eq. (5.26). The size dependence of the magnetization is a bit

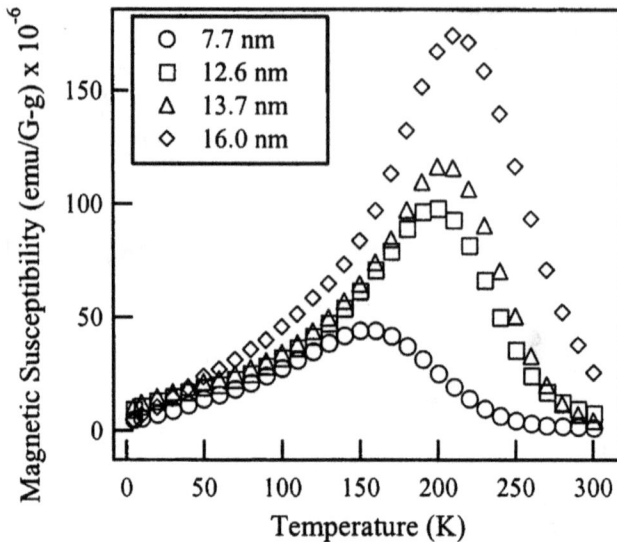

Figure 5.15: Dependence of ZFC curves on particle size for the $CoFe_2O_4$ system of Ref. [4]. Adapted with permission from [4]. © American Chemical Society.

more complicated. Starting from the potential shown in Figure 5.11, Stoner [5] showed that the magnetization of an ensemble of superparamagnetic nanoparticles is given by:

$$M = \mu n L(x),$$

$$L(x) = \coth(x) - \frac{1}{x},$$

$$x = \frac{\mu H}{kT}, \tag{5.27}$$

where μ is the moment of a particle (i.e., the saturation magnetization of an isolated particle, which in turn is proportional to its volume), n is the density of the particles, and the other symbols have the usual meaning. The function $L(x)$ is plotted in Figure 5.16, and it is key to understanding the particle size dependence. First, we note that $L(x)$ increases with increasing x. Let us now assume that H and T are fixed, so that $L = L(\mu)$. L increases with increasing μ, that is, it increases with particle size.

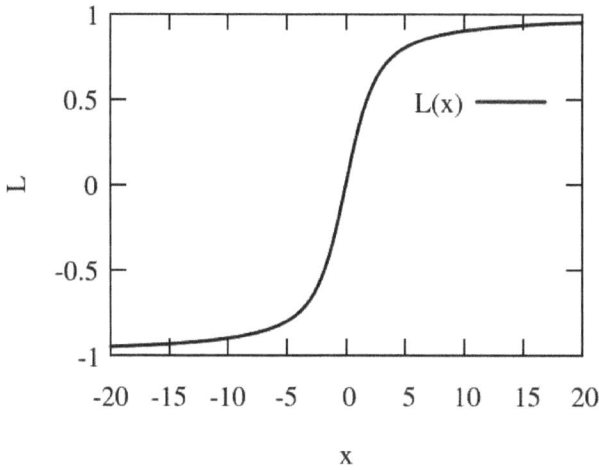

Figure 5.16:   The function $L(x) = coth(1/x) - 1/x$ which is used to calculate the magnetization of superparamagnetic nanoparticles, as in Eq. (5.27).

5.3.4. *Increasing the energy product of magnets*

For permanent magnet applications, an important quantity is the area of the hysteresis curve. This area represents the energy that can be stored in a magnet. A proxy for the area under the curve is the product of saturation magnetization (M_s) and the coercive field (H_c). Given the importance of permanent magnets, considerable research focuses on finding new compounds with improved M_s H_c product. This research is not strictly nanotechnology, of course. However, nanotechnology can be used to improve the stored energy by coupling a soft magnet with high M_s and a hard magnet with (comparatively) low M_s but high H_c. For a series of reasons which go beyond the scope of this book [1], it turns out that the region close ($\sim$3 to $\sim$40 nm, depending on the system) to the interface between a soft and a hard magnet behaves as if it were a hard magnet. This means that the coercivity of this region will be close (sometimes, even higher) to that of the hard magnet. At the same time, this phase will have a high M_s (it is the soft phase), which will increase the magnetization of the composite. An example is shown in Figure 5.17, where SmCo (hard) and FeNi (soft) particles were ball milled to mix

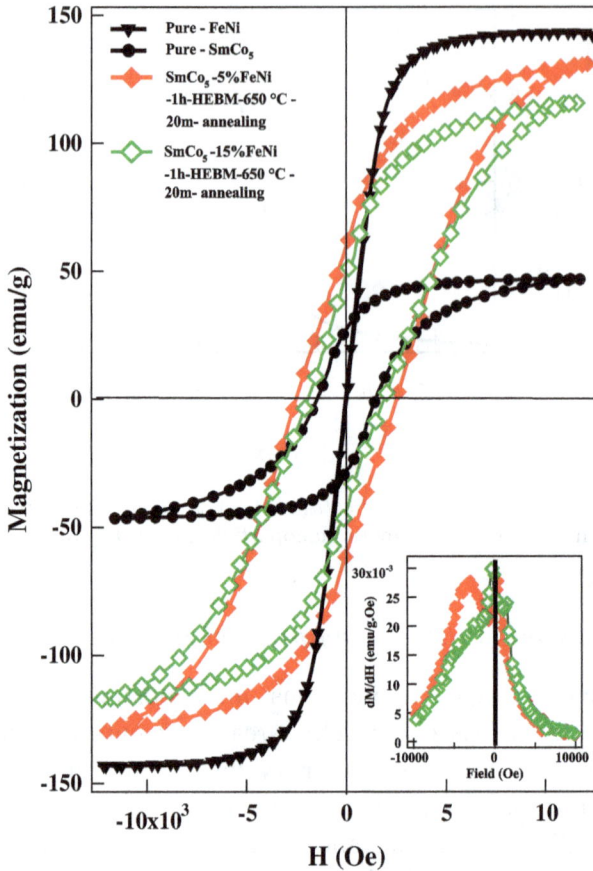

Figure 5.17: Increase of the energy stored inside a magnet by combining a soft and a hard magnetic material. Adapted with permission from [6]. © Elsevier.

the phases [6]. The FeNi phase behaves almost like a superparamagnet, with negligible coercivity but high M_s. The pure SmCo phase presents a high coercivity but its M_s is comparatively low. The composite has a coercivity which is a bit higher than that of the SmCo phase [7] and a M_s intermediate between that of the two materials. This coupling increases the energy product. Let us now focus on the inset of Figure 5.17. It reports dM/dH, which is a useful quantity for this type of work. We know that for a given material dM/dH has a maximum close to its H_c (just look at the hysteresis curves of the

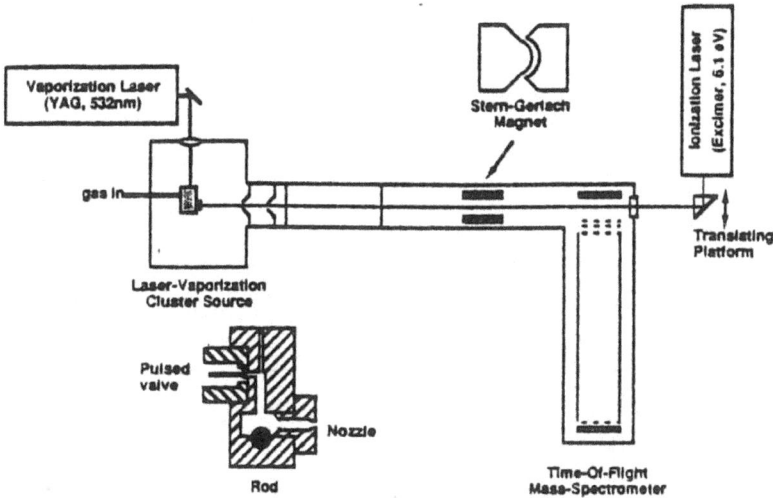

Figure 5.18: Experimental set-up for the measurement of magnetism of metal clusters. Adapted with permission from [8]. © World Scientific.

soft and the hard phase to convince yourself). The inset shows that dM/dH has two maxima. This means that we have multiple phases (that is, incomplete mixing), or that the soft phase is too thick.

5.4. Magnetism of atomic clusters

The question of magnetism of atomic clusters was addressed by seminal experiments from the group of de Heer [8]. In these experiments, a cluster beam was produced and passed through a Stern-Gerlach magnet, as shown in Figure 5.18. The wedged magnet creates a strong magnetic gradient and therefore increases the force on the magnetic clusters: $F \propto \mu \times \nabla B$, where μ is the magnetic moment of the cluster. Results for Ni, Co and Fe are reported in Figure 5.19. The data shows that the magnetic moment decreases from that of a single atom to the bulk value for aggregates of a few hundred atoms. This is consistent with the trends observed for the electronic properties of transition metals reported in Chapter 3.

The magnetization follows the general trend of bulk magnetization when the cluster temperature is varied, as shown in Figure 5.20.

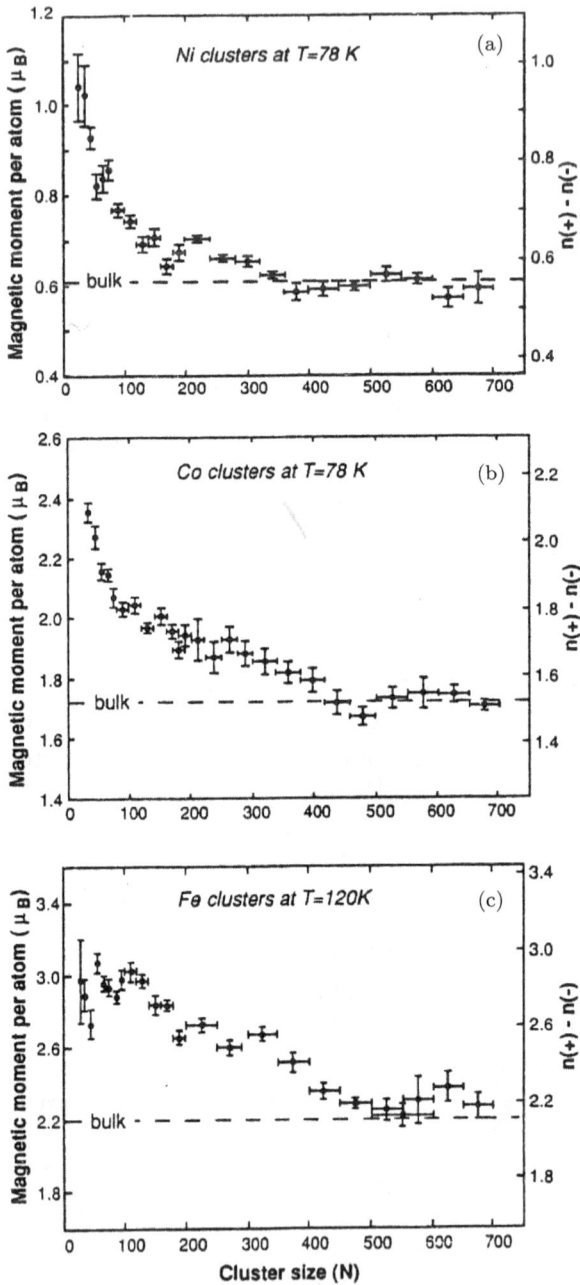

Figure 5.19: Variation of the magnetic moment of metal clusters as function of their size. (a) = Ni, (b) = Co, (c) = Fe. Adapted with permission from [8]. © World Scientific.

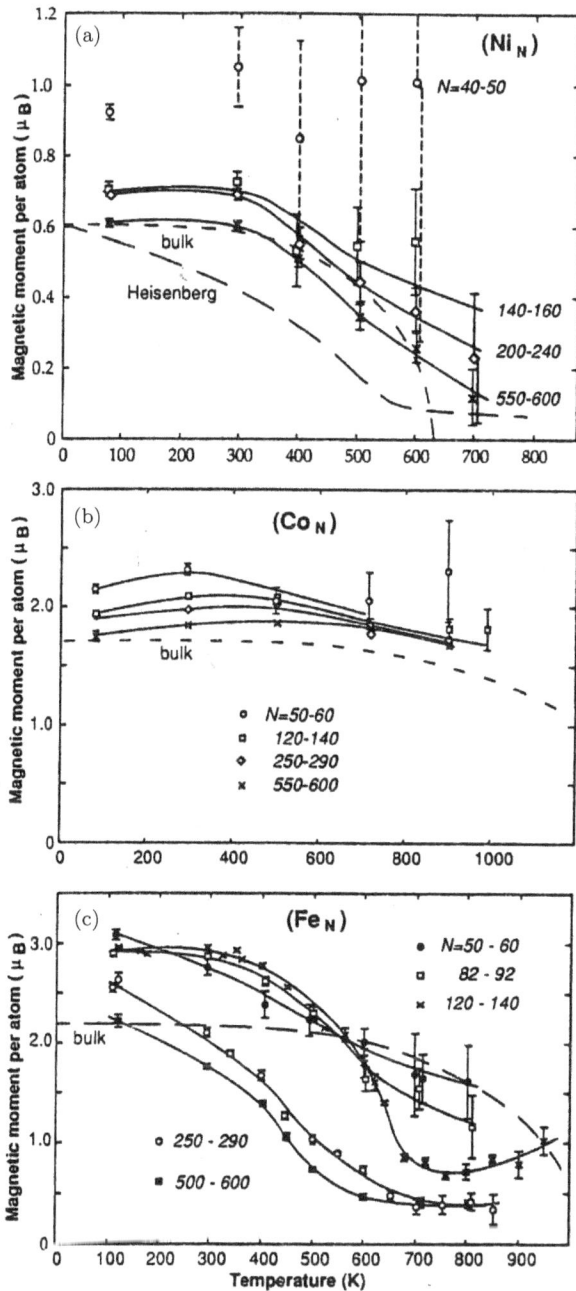

Figure 5.20: Dependence of magnetization on cluster temperature. The dashed line indicates the bulk behavior. Adapted with permission from [8]. © World Scientific.

The Fe data represents an exception, since Fe has a bcc to fcc phase transition at 700 K.

References

1. R. C. O'Handley, "Modern Magnetic Materials, Principles and Applications", Wiley, New York, 2000. Chapter 2.
2. E. P. Wohlfarth, *J. Magn. Magn. Mater* **39**, 39–44 (1983).
3. S. Kamali, C. J. Chen, B. Bates, C. E. Johnson and R. K. Chiang, *J. Phys.: Condens. Matter* **32**, 015302 (2020).
4. C. R. Vestal, Z. J. Zhang, *Chem. Mater.* **14**, 3817–3822 (2002).
5. E. C. Stoner, *Phil. Trans. Roy. Soc. A* **235**, 165 (1936).
6. B. K. Rai, S. R. Mishra *J. Magn. Magn. Mater* **344**, 211–216 (2013).
7. B. Balasubramanian, P. Mukherjee, R. Skomski, P. Manchanda, B. Das and D. J. Sellmyer, *Scientific Reports* **4**, 6265 (2014).
8. I. M. L. Billas, A. Châtelain, W. A. de Heer *Surface Review and Letters*, Vol. 3, No. 1 (1996), pp. 429–434.

Chapter 6

Catalysis

First, the "usual" word of caution. Catalysis has always been "nano". Why? Mostly because in catalysis we need a high surface area. In addition, it was realized early on that metal atoms at edges and corners of nanoparticles will be poorly coordinated and therefore more chemically reactive. This idea was proven in a seminal work by the group of Somorjai, shown in Figure 6.1 [1]. In these experiments, a molecular beam containing a mixture of H_2 and D_2 was directed onto a stepped Pt surface. The reactants dissociated on the Pt surface (the catalyst). The reaction product, HD, was measured with a quadrupole mass spectrometer. The Pt crystal was rotated so that the collimated incident beam would strike perpendicular to the step edges (curve (a)) or parallel to them (curve (b)). The angle of incidence θ was also changed, as shown in Figure 6.1. Curve (c) shows that the reaction probability depends strongly on θ and is highest when the reactants strike the open side of the step structure, as shown in the right lower panel of Figure 6.1. Curve (c) shows no strong dependence on the angle of incidence, as expected for a flat surface. These results show that the step edges, where surface atoms are less coordinated, are more reactive than flat terraces.

6.1. Surface structure

The work by Somorjai is a good starting point to discuss the structure of solid surfaces. Speaking very broadly, we have three main categories of surfaces: flat, stepped and kinked. The structure of these

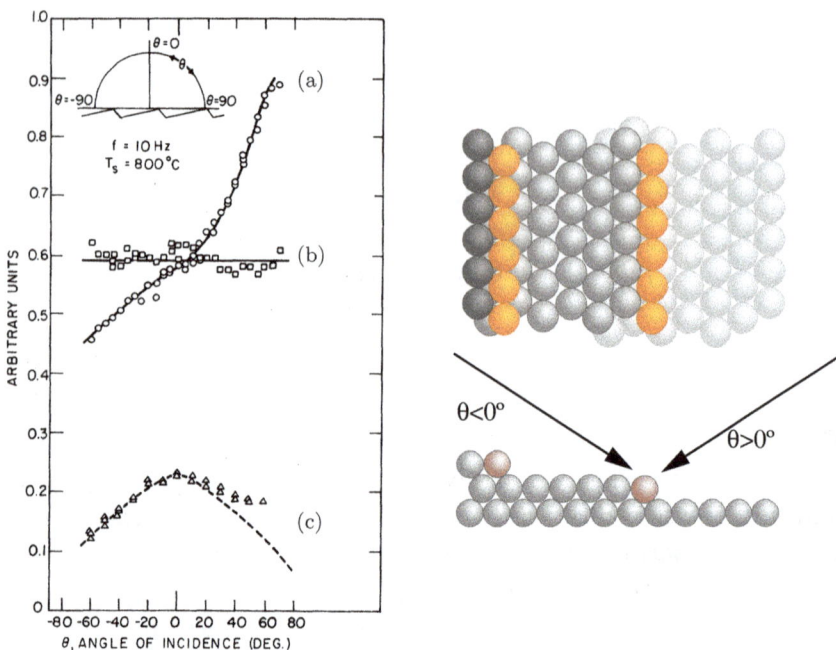

Figure 6.1: Left panel: experimental data showing the higher reactivity of step edges. (a) Incident beam perpendicular to the steps. (b) Incident beam parallel to the steps. (c) Flat Pt surface. Right panel, top: top view of a stepped surface. Lower panel: side view, showing the geometry of curve (a) for different angles of incidence. Adapted with permission from [1]. © American Physical Society.

surfaces can be calculated from the planes defined by Miller indices. Today, a quick internet search will yield the precise arrangement of the atoms on these surfaces, so their derivation will be omitted.

6.1.1. *Flat surfaces*

For fcc crystals, the (111) surface has a cannonball stacking. It is the flattest surface and its atoms have the highest coordination. The (100) surface is quite comparable to the (111), but already the (110) shows atoms with low coordination, as shown in Figure 6.2.

A simplified model of a spherical nanoparticle, shown in Figure 6.3, shows that flat surfaces can be (and, actually, are) encountered in catalysis.

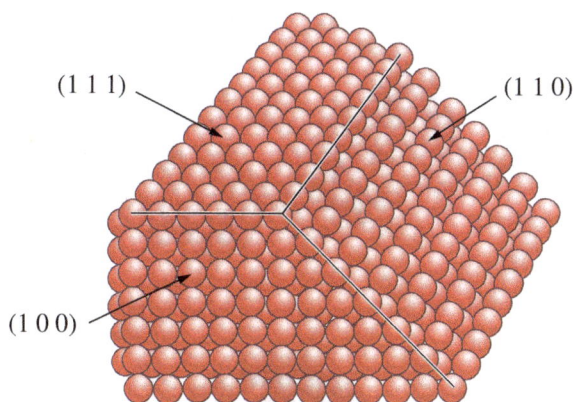

fcc lattice : different net planes

Figure 6.2: Flat surfaces in fcc crystals.

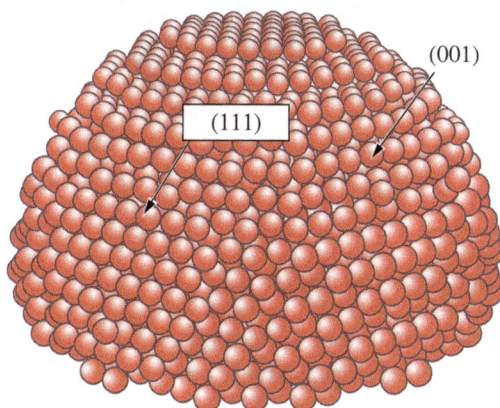

fcc crystal : spherical tip

Figure 6.3: Idealized model of a spherical nanoparticle, showing that the surfaces are not necessarily composed of poorly coordinated atoms.

6.1.2. *Stepped surfaces*

When fcc crystals are cut perpendicular to high index vectors, stepped or even kinked surfaces ensue, such as the one used by Somorjai. Examples are shown in Figure 6.4. Such stepped and kinked surfaces can also be found in nanoparticles and are expected to be most reactive.

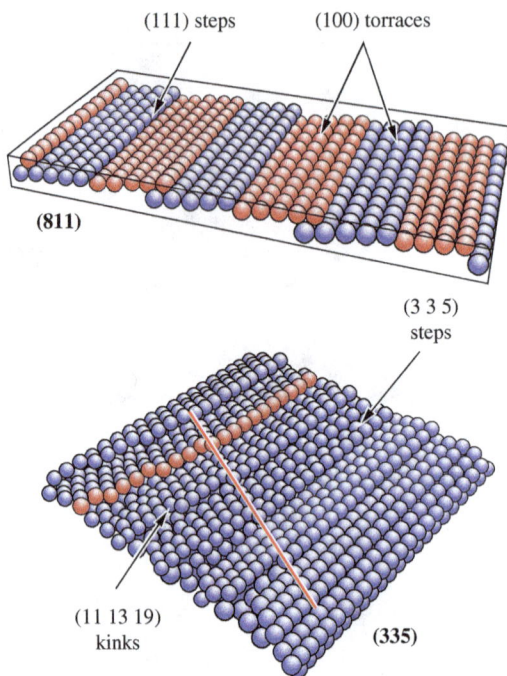

Figure 6.4:   Top: Stepped (811) surface. Bottom: kinked (335) surface.

6.1.3. *Surface reconstruction*

Surfaces are inherently unstable systems. Surface atoms have low coordination and can comparatively easily rearrange their packing. In some cases, surfaces reconstruct spontaneously, even in vacuum. This is, for example, the case of the Au(111) and of the Pt(100) surfaces. A comparison of the atomic arrangement of low-index Au surfaces (theoretical vs. actual, reconstructed) is presented in Figure 6.5.

In many cases, adsorption of molecules induces reconstruction. Sadly, in most cases we cannot anticipate if an adsorbate will induce reconstruction, and how the reconstructed surface will look like. Figure 6.6 reports examples of rather bizarre reconstructions induced by Br adsorption on Cu(100) [2] and Cu(210) [3].

(a) Top views

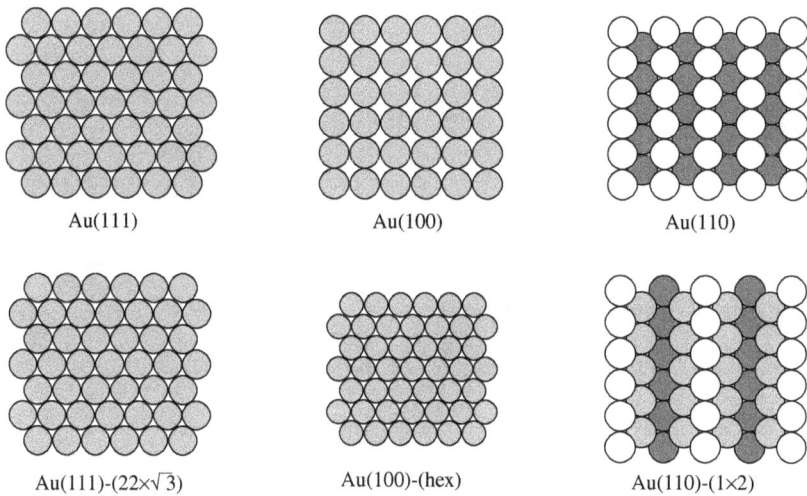

Au(111)

Au(100)

Au(110)

Au(111)-(22×√3)

Au(100)-(hex)

Au(110)-(1×2)

(b) Profile views

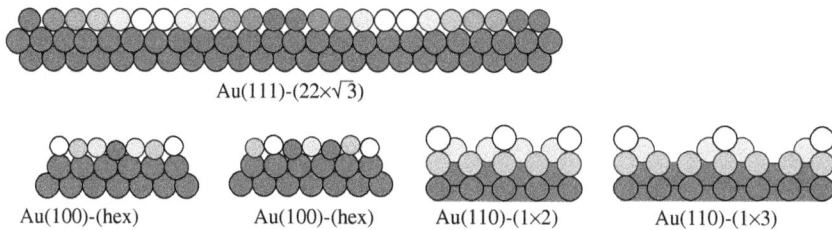

Au(111)-(22×√3)

Au(100)-(hex)

Au(100)-(hex)

Au(110)-(1×2)

Au(110)-(1×3)

Figure 6.5: Top and side views of unreconstructed and reconstructed low-index Au surfaces.

6.2. Reactions at surfaces

Surface chemical reactions are incredibly complex matters. A model that applies to some simple cases (e.g., adsorption of CO on metal surfaces) will be first presented, followed by an analysis of more realistic cases.

Figure 6.6: Reconstruction induced by Br adsorption on Cu surfaces. Adapted with permission from [2, 3]. © Elsevier.

6.2.1. *The adsorption isotherm*

The key to a chemical reaction on a nanoparticle (or, for that matter, on any surface) is that the reactants must reach the surface, adhere (stick) to it, react and then desorb as products. This section will describe a simplified yet powerful model of this ballet. It will be assumed that reagents come onto a surface from the gas phase. The model works just as well for the liquid phase, but there one has to deal with the solvent, which complicates the picture. The assumptions are as follows:

1. Adsorbates do not diffuse after adsorption and occupy only one surface site.
2. The adsorbates do not interact with each other.
3. The maximum coverage is one monolayer, that is, molecules cannot be on top of each other. The coverage is indicated by the letter Θ and is a fraction, that is, one monolayer corresponds to $\Theta = 1$.

The rate of adsorption is proportional to the flux F of molecules to the surface. From kinetic gas theory:

$$F = \frac{1}{4}\frac{n}{V}\bar{v}, \qquad (6.1)$$

where n/V indicates the density of molecules in the gas phase, $\bar{v} = \sqrt{8RT/\pi M}$ is the mean velocity of the molecules in the gas phase, T is the temperature and M is the molecular weight. Remembering that $PV = nRT$, we can write the adsorption rate as:

$$A(\Theta) = \frac{P}{\sqrt{2\pi mkT}} A_0 e^{-\frac{\Delta E_A}{kT}} f(\Theta), \tag{6.2}$$

where A_0 is a proportionality constant, k the Boltzmann constant, m the mass of the molecules, $f(\Theta)$ is the number of available sites and ΔE_A is the activation barrier for adsorption. The activation barrier is negligible for several catalytically relevant transition metals (e.g., H_2 on bare Pd surfaces) but it can be substantial for some systems (e.g., H_2 on noble metals and many oxides). For non-dissociative adsorption (e.g., CO on most transition metals), $f(\Theta) = (1 - \Theta)$. For dissociative adsorption, the fragments of the incident molecule must find free space. In the most simple case where two fragments are generated (e.g., H_2 on Pd), $f(\Theta) = (1 - \Theta)^2$. As in all chemical processes, the reverse reaction, desorption, must also to be taken into account:

$$D(\Theta) = D_0 e^{-\frac{\Delta E_D}{kT}} \Theta \,(\text{non-dissociative case})$$
$$D(\Theta) = D_0 e^{-\frac{\Delta E_D}{kT}} \Theta^2 \,(\text{dissociative case}) \tag{6.3}$$

where D_0 is a proportionality constant and ΔE_D is the activation barrier for desorption.

At equilibrium, $A(\Theta) = D(\Theta)$, and after some rearrangement we obtain:

$$\Theta = \frac{b(T) \cdot P}{1 + b(T) \cdot P} \,(\text{non-dissociative case})$$
$$\Theta = \frac{\sqrt{b(T) \cdot P}}{1 + \sqrt{b(T) \cdot P}} \,(\text{dissociative case}), \tag{6.4}$$

where $b(T)$ is a temperature-dependent factor that takes into account the different activation energies and frequency factors of

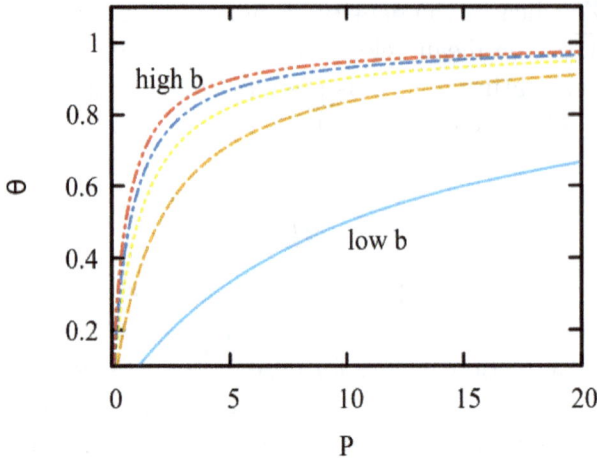

Figure 6.7: Examples of Langmuir adsorption isotherms for different values of $b(T)$.

adsorption and desorption. Examples of Langmuir adsorption isotherms are reported in Figure 6.7. At high pressures, an asymptotic behavior is apparent, which arises from the form of Eq. (6.4). The asymptotic behavior can also be understood by considering the physics of the problem: at high pressures, all the sites are occupied, $\Theta \sim 1$ and not much happens.

Low pressures are a more interesting case:

$$\Theta \approx bP \, (\text{non-dissociative case})$$
$$\Theta \approx (bP)^{1/2} (\text{dissociative case})$$

$$(6.5)$$

that is, the type of adsorption (non-dissociative vs. dissociative) can be distinguished by looking at the pressure dependence for low coverages.

In most real-world cases, molecule (atom) A and molecule B will stick to the surface, dissociate if necessary, diffuse and then react. Thus, a surface reaction is a pretty complicated ballet, whose outcome is quite difficult to predict. Some quantities, such as the sticking coefficient of simple molecules at low coverages, are comparatively easy to measure and can be simulated by theory relatively well. However, things become quite complicated when polyatomic molecules are involved, or when the coverage is non-negligible. For example, on many metals, O_2 goes first through a weakly chemisorbed state before dissociating.

6.3. Model systems

6.3.1. *Selected hydrocarbon reactions*

The outcome of a reaction is anything but unique, and hard to anticipate. See, for example, Figure 6.8 for a quite "trivial" reaction on Pt(111), ethylene hydrogenation. In this reaction, ethylene and hydrogen are passed on a Pt surface, yielding ethane. The top panel of Figure 6.8 is an idealized representation of the reaction, which proceeds through the following steps:

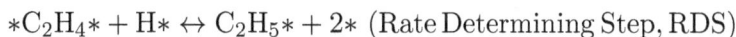

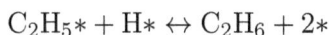

$H_2 + 2* \leftrightarrow 2H*$

$C_2H_4 + * \leftrightarrow C_2H_4*(\pi - \text{bonded})$

$C_2H_4* + * \leftrightarrow *C_2H_4*(\text{di} - \sigma \text{ bonded})$

$*C_2H_4* + H* \leftrightarrow C_2H_5* + 2*$ (Rate Determining Step, RDS)

$C_2H_5* + H* \leftrightarrow C_2H_6 + 2*$

$C_2H_4 + H_2 \leftrightarrow C_2H_6(\text{sum})$,

where * indicates a surface site. In the ideal model, hydrogen dissociates on the surface and diffuses to react with an adsorbed ethylene molecule. Since the molecule is di-σ bonded to the surface, addition of hydrogen is the rate-determining step. The bottom panel of Figure 6.8 reports vibrational spectroscopy analysis of the adsorbed species. Some of the hydrocarbon peaks are consistent with the predictions of the model. However, ethylidyne is also detected [4]. Ethylidyne is a dehydrogenated hydrocarbon which is attached very strongly to the substrate. It requires three hydrogen atoms to react with it before it can desorb. This is of course unlikely to happen. The most likely destiny of the ethylidyne fragment is to remain adsorbed on the surface, limit the number of available surface sites, and reduce the reactivity. Worse yet, the fragment can further decompose, for example loose additional hydrogen atoms, and yield, in the end, a layer of carbon-rich fragments or even elemental carbon, all strongly bound to the surface. When a large number of these fragments forms,

(a)   (b)   (c)

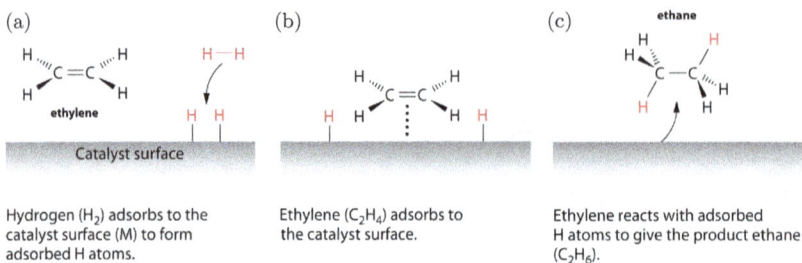

Hydrogen (H$_2$) adsorbs to the
catalyst surface (M) to form
adsorbed H atoms.

Ethylene (C$_2$H$_4$) adsorbs to
the catalyst surface.

Ethylene reacts with adsorbed
H atoms to give the product ethane
(C$_2$H$_6$).

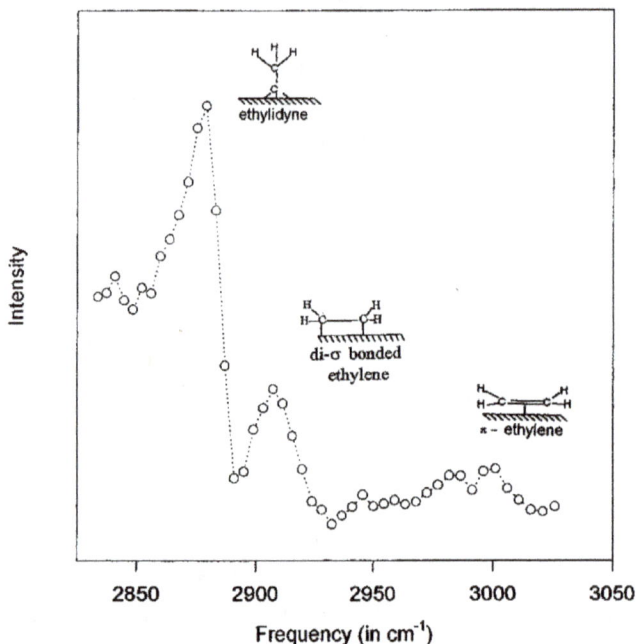

Figure 6.8: Hydrogenation of ethylene. Top panel: ideal process. Bottom panel: vibrational spectroscopy analysis, showing a multitude of adsorbed hydrocarbon species. Bottom panel adapted with permission from [4]; © American Chemical Society.

the catalyst must be removed from the reactor and regenerated. For ethylene hydrogenation, nanotechnology is relevant to control catalyst size. For example, for Pt particles supported on SiO$_2$ and Al$_2$O$_3$ with a size above ~2 nm the activity is comparable to that of single-crystal surfaces. The activity increases for particles with size between ~2 nm and 0.6 nm and becomes negligible below about 0.6 nm [5]. The reasons for this strong size dependence have yet to

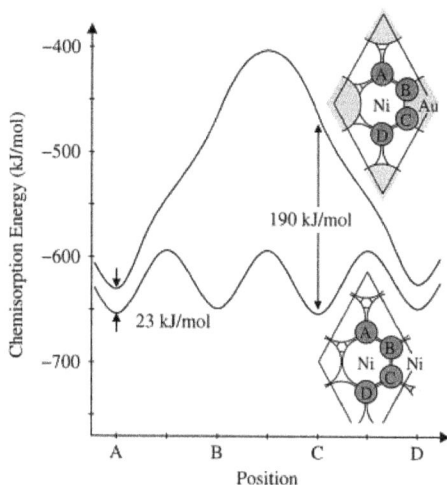

Figure 6.9:   Bottom curve: adsorption energy of carbon on a pure Ni(111) surface. Note how the energy does not depend on position. Top curve: adsorption energy of carbon on a Ni-Au surface. Note the strong dependence on position and the 30% reduction in adsorption energy in the sites closer to the Au atom. Adapted with permission from [6]. © AAAS.

be explained. Yet, particle size control is extremely relevant for this system.

Accumulation of carbonaceous compounds on the surface can be alleviated by adding non-reactive metals, such as Au, to a surface. In Ref. [6] theoretical calculations were coupled with experiment to show that addition of Au to a Ni surface decreased the adsorption energy of carbon, thereby facilitating its desorption. Figure 6.9 shows that addition of Au can reduce the adsorption energy by up to 30% compared to the bare Ni surface. Correspondingly, a Ni-Au catalyst is considerably more resiliant to poisoning than the bare Ni for the steam reforming of n-butane, as shown in Figure 6.10.

6.3.2. *Hydrodesulfurization*

Sulfur-containing hydrocarbons are a nuisance for the petroleum and energy industries. If present in the feed, they tend to poison transition metal catalysts (e.g., Pt) which are commonly employed in the petrochemical industry. If present in the exhaust of a process, sulfur-containing compounds cause environmental and health issues.

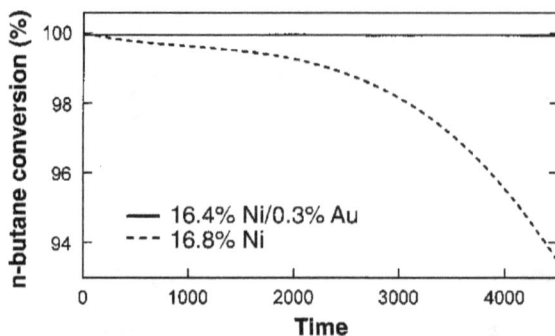

Figure 6.10: Conversion of n-butane as a function of time during steam reforming. Adapted with permission from [6]. © AAAS.

Hydrodesulfurization (HDS) is the reaction where hydrogen and sulfur-containing hydrocarbons are reacted to yield the volatile H_2S, which can then be removed by an amine scrubber.

The most popular HDS catalyst is Mo, supported on Al_2O_3 and aided by Co as a promoter. The mechanism of action of this CoMo catalyst has been discussed for a long time. Initially, the general idea was that Mo had a high affinity for S and therefore it reacted with S to yield MoS_2. In presence of excess H_2, it was thought that surface S would react to form H_2S. The formation of the surface sulfur vacancy was thought to be accompanied by the reduction of the metal ion from Mo^{4+} to Mo^{3+}. The Mo^{3+} ion would then react with an incoming sulfur-containing molecule and oxidize again to Mo^{4+}.

Roughly:

$$Mo^{+4}S_2 + H_2 \rightleftarrows Mo^{+3}S^+ + H_2S^-$$

$$Mo^{+3}S^+ + RS \rightleftarrows Mo^{+4}S_2 + RS^+.$$

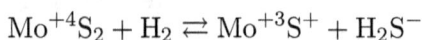

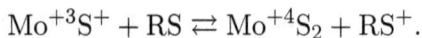

Starting in the early 2000s, atomically resolved STM studies showed that the mechanism was much more complex than originally thought [7–9]. The key finding was that small MoS_2 clusters possessed a one-dimensional metallic state that followed the rim of the nanoparticles, as shown in Figure 6.11(a) [10]. The same set of studies also determined that the flat part of the nanoparticles was not

(a)   (b)

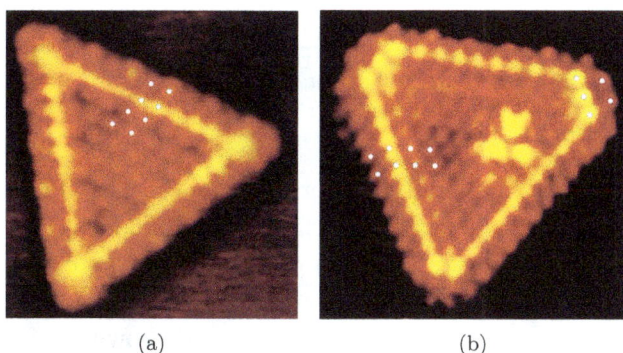

Figure 6.11: (a): STM image of a MoS$_2$ cluster deposited on a Au(111) surface. The side of the triangle is about 5 nm. The bright yellow contour is the signature of a one-dimensional metallic state. (b): same, with Co added as a promoter. Adapted with permission from [10]. © Academic Press.

chemically active. Instead, the metallic state was responsible for the chemical activity of the catalyst. Since the rim behaved like a metal, hydrogen atoms would readily stick to it and react with sulphur-containing molecules in the feed. STM also clarified the role of the promoters. Addition of Co to a MoS$_2$ nanoparticle changed the morphology to a truncated triangle, as shown in Figure 6.11(b). Some of the truncated triangular prisms appeared to have rounded corners, indicating that high-index edge terminations were exposed. The atoms in these high-index terminations had poor coordination and were more reactive. The same studies also showed that the support played an unforeseen role. Until then, Al$_2$O$_3$ was the most popular support for HDS catalysts because of its low cost and ease of preparing supports with high surface area. However, the MoS$_2$ catalyst interacts strongly with the oxygen atoms of the support. Mo-O-Al bonds form. Since the catalyst consists of one atomic layer, these bonds can significantly alter its electronic structure and performance. It was found that desulfurization at high temperatures eliminated bonding to the substrate and increased the reactivity. However, high temperature processing led to coalescence of the catalyst nanoparticles, which is undesirable in catalysis. Therefore, it was suggested that substrates interacting weakly with Mo should be used *in lieu* of Al$_2$O$_3$. This research led to the development of two new HDS catalysts, marketed by Topsøe under the brand name TK-558 BRIM and TK-559 BRIM (both names are trademarks).

6.3.3. *Ceria supports*

Supports are of paramount importance in catalysis. They anchor the catalyst nanoparticles (especially metallic ones) and prevent their sintering, but they can also contribute to their activity. One of the most interesting catalyst supports is cerium oxide (ceria), CeO_2. CeO_2 has a cubic fluorite structure, shown in Figure 6.12 [11]. It is a peculiar material, since Ce^{4+} can be readily reduced to Ce^{3+}. In fact, the electrochemical reaction in aqueous solution, $Ce^{4+} + e^- \rightleftharpoons Ce^{3+}$, has a quite high potential ($E^0 = 1.44\,V$ vs. NHE). Reduction of Ce^{4+} to Ce^{3+} yields two defect types: interstitial oxygen and oxygen vacancy (which, however, requires two Ce^{4+} ions to be reduced), but does not change the crystalline structure, as shown in Figure 6.13 [12]. Overall, the high reduction potential of Ce^{4+} means that oxygen in ceria is more reactive than in other oxides. This high oxygen reactivity makes ceria an excellent catalyst support; metal nanoparticles tend to adhere more strongly, and to be more finely dispersed, to ceria than to other oxides. The oxygen reactivity also makes ceria a promoter for the water gas shift reaction, as shown in Figure 6.14:

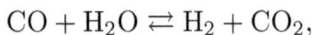

$$CO + H_2O \rightleftharpoons H_2 + CO_2,$$

and helps prevent formation of carbonaceous deposits in the CO_2 – methane reforming reaction, as shown in Figure 6.15 [13]:

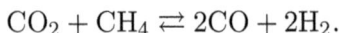

$$CO_2 + CH_4 \rightleftharpoons 2CO + 2H_2.$$

Let us now consider low-index surfaces, which are the most likely to be exposed in ceria nanoparticles. These surfaces are the (111), the (110) and the (100), which are shown in Figure 6.16 [14]. The most stable surface is the (111), where Ce is coordinated to 7 O atoms and O is coordinated to 3 Ce atoms. In the (110) plane, Ce is coordinated to 6 O atoms and O is coordinated to 3 Ce atoms. In the (100) plane, Ce is coordinated to 6 O atoms and O is coordinated to 2 Ce atoms.

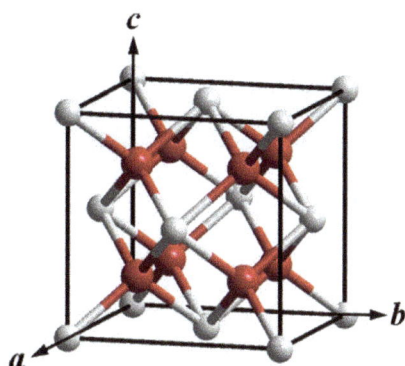

Figure 6.12: Structure of CeO$_2$. Ce: white spheres. O: red spheres. Adapted with permission from [11]; © Royal Society of Chemistry.

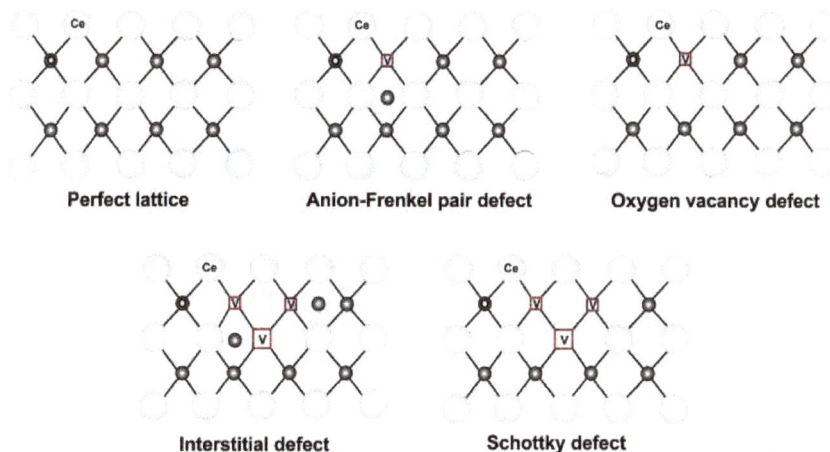

Figure 6.13: Right: perfect CeO$_2$ lattice. Center: Interstitial oxygen defect. Left: Oxygen vacancy defect. Adapted with permission from [12]. © Elsevier.

The lower coordination makes the (100) plane less stable than the (111) and (110) planes. Recent work by the group of Datye [15] has focused on the structure and reactivity of nearly monodisperse Ceria nanoparticles with well-defined shapes. Octahedra, rods and fibers were first analyzed by aberration-corrected, high-resolution TEM. Representative bright field images are reported in Figure 6.17 for octahedral nanoparticles, and they show a prevalence of the (111)

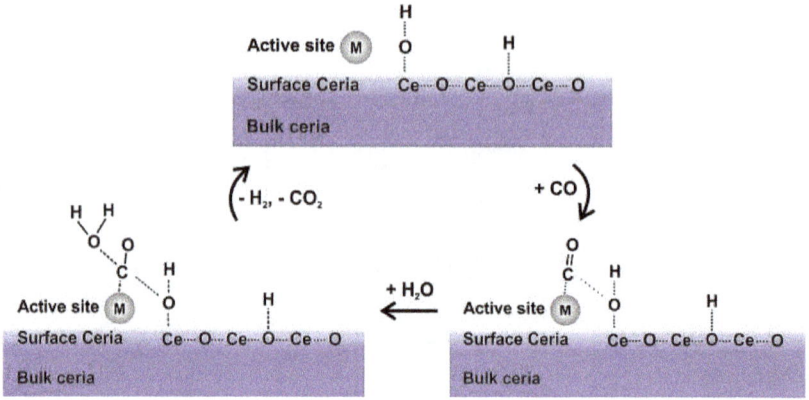

Figure 6.14: Ceria as promoter of the water gas shift reaction. Adapted with permission from [12]. © Elsevier.

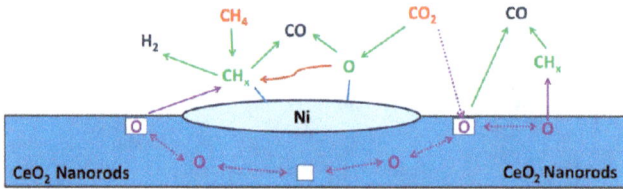

Figure 6.15: Oxygen defects in ceria help prevent formation of carbonaceous deposits in the methane reforming reaction. Adapted with permission from [13]; © American Chemical Society.

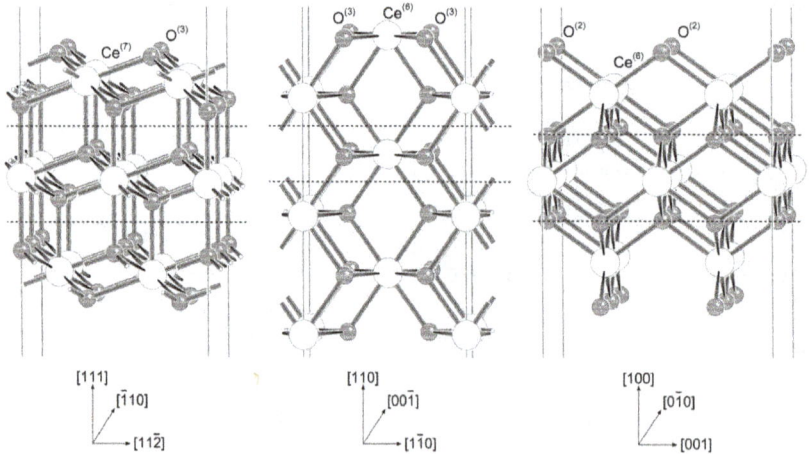

Figure 6.16: Low-index surfaces of CeO_2. Adapted with permission from [14]; © Elsevier.

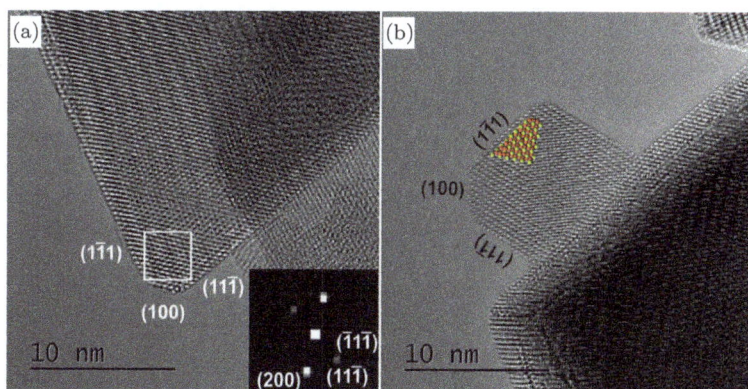

Figure 6.17: High resolution, aberration-corrected TEM of octahedral nanoparticles. Note the prevalence of (111) surfaces. (100) can be found only at the edges. Adapted with permission from [15]; © Wiley-VCH Verlag GmbH & Co. KGaA, Weinheim.

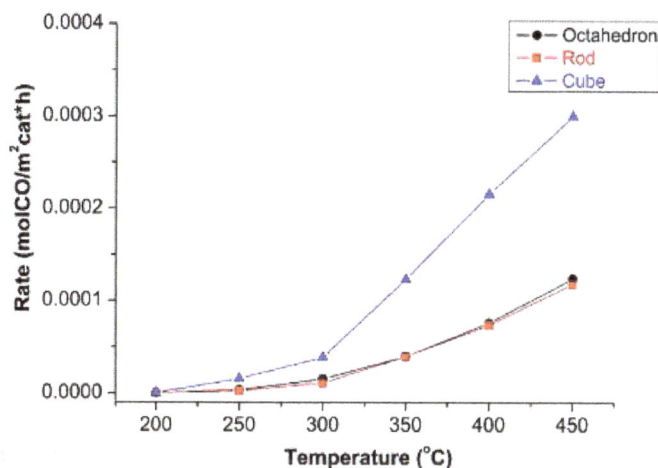

Figure 6.18: Rate of CO conversion for the water gas shirt reaction for different shapes of the ceria catalyst. Adapted with permission from [12]. © Elsevier.

surface. Rods also presented mostly (111) surfaces, while cubes were prevalently (100).

The ceria nanoparticles were then tested as catalysts for the water-gas shift reaction. Results are presented in Figure 6.18, expressed in moles of CO product per square meter of CeO_2 catalyst.

This representation allows to correlate structure and activity. The conversion rates for octahedra and rods are comparable, which is not surprising (the nanoparticles expose the same (111) facets). Cubes, instead, present a much higher activity, due to the presence of the more reactive (100) surfaces.

References

1. R. G. Gale, M. Salmeron, and G. A. Somorjai, *Phys. Rev. Lett.* **38**, 1027–1029 (1977).
2. T. W. Fishlock, J. B. Pethica, R. G. Edgell, *Surf. Sci.* **445**, L47 (2000).
3. A. T. S. Wee, T. W. Fishlock, R. A. Dixon, J. S. Foord, R. G. Edgell, J. B. Pethica, *Chem. Phys. Lett.* **298**, 146 (1998).
4. P. S. Cremer, X. Su, Y. R. Shen, G. A. Somorjai, *J. Am. Chem. Soc*, **118**, 2942–2949 (1996).
5. C. R. Henry, *Surface Science Reports*, **31**, 231 (1998).
6. F. Besenbacher, I. Chorkendorff, B. S. Clausen, B. Hammer, A. M. Molenbroek, J. K. Nørskov, I. Stensgaard Science 279, 1913 (1998).
7. M. V. Bollinger, J. V. Lauritsen, K. W. Jacobsen, J. K. Nørskov, S. Helveg, and F. Besenbacher, *Physical Review Letters* **87**, 196803 (2001).
8. J. V. Lauritsen, M. Nyberg, R. T. Vang, M. V. Bollinger, B. S. Clausen, H. Topsøe, K. W. Jacobsen, E. Lægsgaard, J. K. Nørskov, and F. Besenbacher *Nanotechnology* **14**, 385–389 (2003).
9. F. Besenbacher, M. Brorson, B. S. Clausen, S. Helveg, B. Hinnemann, J. Kibsgaard, J. V. Lauritsen, P. G. Moses, J. K. Nørskov, H. Topsøe, *Journal of Catalysis* **224**, 94–106 (2004).
10. J. V. Lauritsen, S. Helveg, E. Lægsgaard, I. Stensgaard, B. S. Clausen, H. Topsøe, F. Besenbacher, *J. Catal* 197, 1–5 (2001).
11. C. Sun, D. Xue, *Phys. Chem. Chem. Phys. 15*, 14414 (2013).
12. Z. Wu, S. H. Overbury (Eds.), "Catalysis by Materials with Well-Defined Structures", Elsevier (2015). see Chapter 2.
13. X. Du, D. Zhang, L. Shi, R. Gao, J. Zhang, *J. Phys. Chem. C* **116**, 10009–10016 (2012).
14. M. V. Ganduglia-Pirovano, A. Hofmann, J. Sauer, *Surf. Sci. Rep.* **62**, 219–270 (2007).
15. S. Agarwal, L. Lefferts, B. L. Mojet, D. A. J. M. Ligthart, E. J. M. Hensen, D. R. G. Mitchell, W. J. Erasmos, B. G. Anderson, E. J. Olivier, J. H. Neethling, A. K. Datye, *ChemSusChem* **6**, 1898–1906 (2013).

Chapter 7

Photonic Crystals

Photonic crystals are assemblies of units (e.g., films, fibers, spheres) where the dielectric function is periodic in one, two, or three spatial dimensions (1D, 2D, 3D), as shown in Figure 7.1.

In this Chapter, we will focus our attention on simple systems composed of transparent materials with a real dielectric function. While the theory for 1-, 2-, and 3-D systems will be presented, particular attention will be devoted to 1D systems. The simplicity of these systems allows the illustration of several more sophisticated physical principles, such as "slow-light", cavity fabrication, and waveguiding.

7.1. Theory

One starts from Maxwell'equations:

$$\nabla \cdot \mathbf{D} = 0 \quad \nabla \times \mathbf{E} = \frac{-\partial \mathbf{B}}{\partial t}$$
$$\nabla \cdot \mathbf{B} = 0 \quad \nabla \times \mathbf{H} = \frac{\partial \mathbf{D}}{\partial t} \tag{7.1}$$

We note that all the fields are assumed to be a function of place and time:

$$\mathbf{D} = \mathbf{D}(\mathbf{r}, t), \text{etc.}$$

We also assume:

$$\mathbf{B} = \mu_0 \mathbf{H}, \text{and } \mathbf{D} = \epsilon_0 \epsilon \mathbf{E}.$$

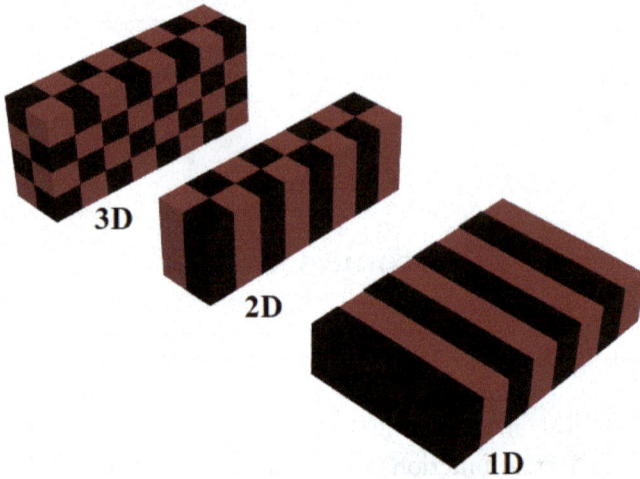

Figure 7.1: Examples of 1D, 2D and 3D photonic crystals.

So far, nothing new. However, one must bear in mind that photonic crystals are not homogeneous materials. Their dielectric function depends on the position and is periodic:

$$\epsilon(\mathbf{r}) = \epsilon(\mathbf{r} + \mathbf{R}), \tag{7.2}$$

where $\mathbf{R}$ is a Bravais lattice vector.

As commonplace in electromagnetic theory, Maxwell's equations are solved by taking the rotation of the field equations:

$$\nabla \times \nabla \times E = -\nabla^2 \mathbf{E}(\mathbf{r}) = \frac{\omega^2}{c^2}\epsilon(\mathbf{r})\mathbf{E}(\mathbf{r})$$

$$\nabla \times \left[\frac{1}{\epsilon(r)}\nabla \times \mathbf{H}\right] = \frac{\omega^2}{c^2}\mathbf{H}(\mathbf{r}) \tag{7.3}$$

These equations have a very useful property for the treatment of photonic crystals: scale invariance. That is, if $\mathbf{E}(\mathbf{r})$ is a solution with a frequency ω for a region with $\epsilon(\mathbf{r})$, then a system with $\epsilon(\mathbf{r}) = \epsilon(\mathbf{r}/s)$, s being a scalar, the corresponding solution is $\mathbf{E}(\mathbf{r}) = \mathbf{E}(\mathbf{r}/s)$ and the frequency is ω/s. We also note that the periodicity of the dielectric constant, Eq. (7.2), is (mathematically speaking) equivalent to the periodicity of the electronic potential in solids. Therefore, the solutions for the electromagnetic field have the form

$$\mathbf{E}(\mathbf{r}) = \mathbf{E}_{k,n}(r)$$
$$\mathbf{E}_{k,n}(\mathbf{r} + \mathbf{R}) = e^{i\mathbf{k}\cdot\mathbf{R}}\mathbf{E}_{k,n}(\mathbf{r}),$$

(7.4)

where n is the band index and $\mathbf{k}$ is the wave vector within the first Brillouin zone. There are several methods for solving Eqs. (7.3), which are beyond the scope of this work. The interested reader can found an excellent introduction, for example, in Ref. [1].

Once a band structure is obtained, the density of states (defined as the number of states between ω and $\omega + d\omega$) is calculated by integrating over the Brillouin Zone (BZ) as follows:

$$\rho(\omega) = \frac{1}{(2\pi)^3} \sum_n \int_{BZ} \delta[\omega - \omega_n(\mathbf{k})]d\mathbf{k}.$$

(7.5)

A related yet more relevant quantity is the local density of states, LDOS. LDOS gauges the interaction between the electromagnetic field and a dipole $\mathbf{p}$ (that is, an emitter, which may be added to a photonic crystal, as we will see later):

$$\rho_l(\omega, \mathbf{r}, \mathbf{p}) = \frac{1}{(2\pi)^3} \sum_n \int_{BZ} |\mathbf{p} \cdot \mathbf{E}_{\mathbf{k},n}| \delta[\omega - \omega_n(\mathbf{k})]d\mathbf{k}.$$

(7.6)

Using Fermi's golden rule, the LDOS determines the radiative decay rate of the emitter:

$$\Gamma = \frac{2\pi}{\hbar}^2 \rho_l(\omega, \mathbf{r}, p),$$

(7.7)

which tells us that a photonic crystal can be used to modify the lifetime ($= 1/\Gamma$) of an emitter. We also note that within a photonic band gap, $\rho = \rho_l = 0$. Therefore, $\Gamma = 0$; hence radiative decay can be dramatically reduced (ideally, completely eliminated) in photonic crystals.

7.2. 1D systems

7.2.1. *A most simple system*

Consider the system of Figure 7.2: a wave with wavelength λ crosses a region of space where semitransparent mirrors are placed at a

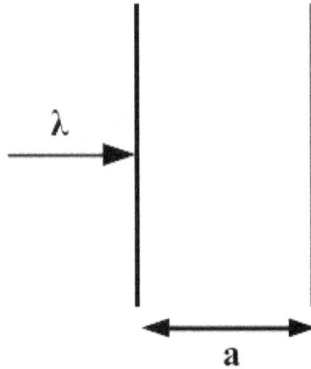

Figure 7.2: The most simple system: A wave incident on mirrors spaced a distance a from each other.

distance a from each other. To further simplify, let us consider only two mirrors. The wave reflected back from the second mirror will interfere destructively with the wave incident on the first mirror when:

$$2a = (m+1)\frac{\lambda}{2}, \quad m \text{ integer}, \tag{7.8}$$

That is, when $a = \lambda/4$. Should we have more than two mirrors, and assuming that the mirrors are semitransparent, part of the incident wave will cross in the region past the second mirror, where, however, it will interfere destructively with the wave reflected back by mirror #3, and so on. We conclude that an appropriate mirror spacing can be used to filter out a wavelength from an incident, polychromatic beam.

7.2.2. *A bit more complicated: A stack of dielectric layers*

Consider the system of Figure 7.3. It consists of a stack of two different materials, with dielectric constants ε_l and, respectively, ε_h, indexes of refraction n_l and, respectively, n_h and thicknesses a_l and, respectively, a_h. We assume the z axis to be perpendicular to the thin films. We saw in Eq. (7.8) that in air (index of refraction $n = 1$) the condition for destructive interference is $a = \lambda/4$. Now, because

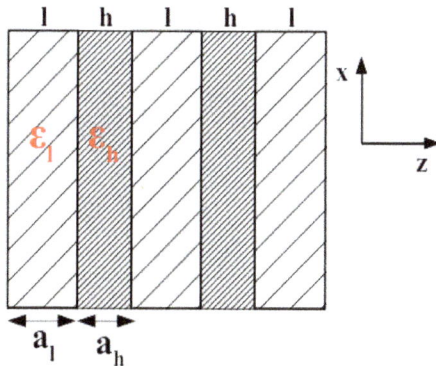

Figure 7.3: Prototype of a 1D photonic crystal.

of the different dielectric functions (that is, indexes of refraction), the wavelength will be different in each layer: $\lambda_{(layer)} = \frac{\lambda_{vacuum}}{n_{layer}}$ The condition for destructive interference then becomes:

$$a_l = \frac{\lambda_{vacuum}}{4n_l}, a_h = \frac{\lambda_{vacuum}}{4n_h}. \tag{7.9}$$

Equation (7.9) shows that selected wavelengths can removed from an incident beam using a suitable combination of materials and thicknesses.

Equation (7.9) is a nice result, but there is more to be learned from the system of Figure 7.3. For this, we need a more rigorous treatment. One possibility is to use a formalism comparable to that of the Kronig-Penney model for electrons in a periodic potential [2]. We assume the field to be incident at an angle, so that it will have x and z components. The electric field is written as

$$E(x, z, t) = E(z)e^{(iqz)}e^{i(k_x x - \omega t)}, \tag{7.10}$$

where k_x is the x-component of the incident wave vector, q is a Bloch (i.e., reciprocal lattice) vector, ω the frequency and the field amplitude is periodic:

$$E(z) = E(z + \Lambda), \tag{7.11}$$

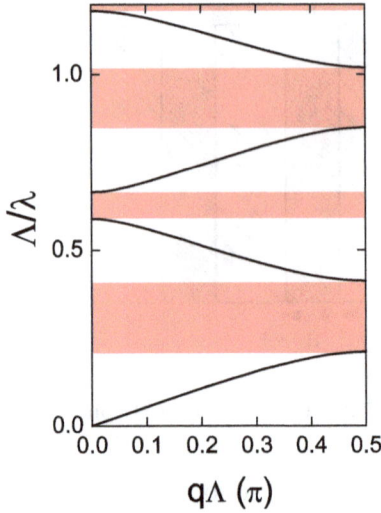

Figure 7.4: Photonic band structure of a 1D system where $a_h/\Lambda = 0.3$, $\varepsilon_h = 9$, $\varepsilon_l = 1$. Adapted with permission from [3]. © Springer.

where $\Lambda = a_l + a_h$. Similar to the Kronig-Penney model, one obtains an implicit equation:

$$\cos(q\Lambda) = \cos(k_{1,z}a_l)\cos(k_{2,z}a_h)$$
$$-\frac{1}{2}\left[\frac{\alpha_1}{\alpha_2} + \frac{\alpha_2}{\alpha_1}\right]\sin(k_{1,z}a_l)\sin(k_{2,z}a_h), \qquad (7.12)$$

where

$$k_{i,z} = \sqrt{\left(\frac{\omega}{c}\right)^2 \epsilon_i - k_x^2},$$

and α_1, α_2 are numerical coefficients that depend on the polarization of the incident light. Like in the Kronig-Penney model, Eq. (7.12) has a solution when the absolute value of its right side is ≤ 1. In that case the wave can propagate through the system.

Figure 7.4 shows the band structure of a 1D system where $a_h/\Lambda = 0.3$, $\varepsilon_h = 9$, $\varepsilon_l = 1$. Band gaps are evident. We also note that the dependency of the energy (wavelength, vertical axis) on the wave vector (horizontal axis) is not linear, depends strongly on the wave vector and is nearly flat at the zone boundaries. This has important consequences. To understand these consequences, one has to introduce the concept of group velocity.

7.2.3. *Group velocity*

Consider a monochromatic wave propagating in vacuum. This wave will have a frequency ω, a wave vector k, and it will propagate with a velocity

$$v_p = \frac{\omega}{k}, \tag{7.13}$$

where the subscript p indicates that this velocity is the *phase velocity*.

In Chapter 4, we also saw that that the relationship between ω and k can be more complicated than the simple relation of Eq. (7.13). A non-linear dispersion curve means that waves with different frequencies will travel through the medium with different velocity. This consideration is key to understand the concept of group velocity.

Let us now consider a wave packet, that is, waves with different wave vector k, traveling through a medium. If the dispersion relation is non-linear, the different waves will propagate with different velocities and the wave packet will change shape while propagating. Hence, the wavepacket will be a function of the position (x) and the time (t): $\alpha(x, t)$. Instead of the wavepacket, it is more convenient to use its Fourier transform:

$$\alpha(x, t) = \int_{\mathbb{R}} A(k) e^{i(kx - \omega t)} dk. \tag{7.14}$$

Let us now assume that the wavepacket is nearly monochromatic:

$$\omega(k) = \omega_0 + (k - k_0)\omega_0', \quad \omega_0' = \frac{\partial \omega}{\partial k}. \tag{7.15}$$

Then:

$$\alpha(x, t) = e^{i(k_0 x - \omega_0 t)} \int_{\mathbb{R}} A(k) e^{i(k - k_0)(x - \omega_0' t)} dk. \tag{7.16}$$

The first term on the right side of Eq. (7.16) is a monochromatic wave that propagates with a velocity given by Eq. (7.13). The second term is an amplitude modulation. This modulation (which is what really counts in signal transmission) will move through the medium with a velocity ω_0'. Thus, the group (= envelope) velocity is

$$v_g = \frac{\partial \omega}{\partial k}. \qquad (7.17)$$

7.2.4. *Slow light*

If we look at the photonic band of Figure 7.4, we note several near-horizontal regions, typically located near the zone edges. In these regions, v_g is small. This is the concept of slow light: the signal propagates with a velocity which can be substantially lower than the speed of light in vacuum. An alternative way of depicting this phenomenon is to state that the index of refraction becomes very big when the group velocity is small:

$$v_g = \frac{c}{n_g}. \qquad (7.18)$$

Slow light generates a number of quite interesting phenomena:

- Pulse compression and intensity enhancement. When an optical pulse enters a slow-light material, the front of the pulse is slowed down. The tail of the pulse travels faster than the front, and the pulse shortens, as shown in Figure 7.5. Since energy is conserved

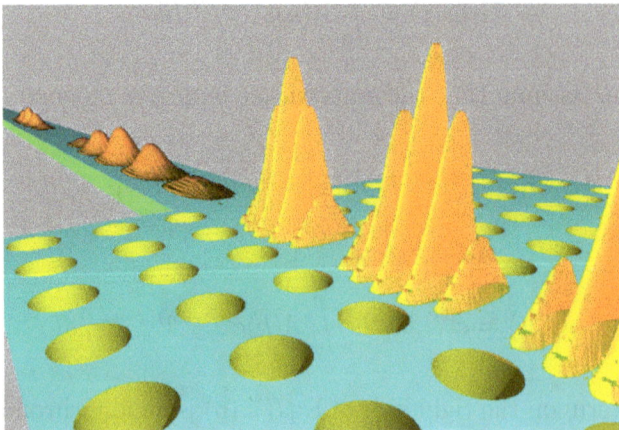

Figure 7.5: Compression of a pulse upon entrance in a slow light material. Adapted with permission from [4]. © IOP Publishing.

in the transition to the slow light region, the shorter pulse also gains intensity [4]:

$$I_{SL} = I_0 S, \text{where } S = \frac{v_p}{v_g} = \frac{n_p}{n_g} \qquad (7.19)$$

• Optical switching. In non-linear optical materials, the index of refraction can be expressed as

$$n = n_0 + n_2 I, \qquad (7.20)$$

where n_0 is the linear index, n_2 the non-linear coefficient and I is the light intensity. Non-linear materials can be used as switches. For example, one can place a non-linear material in one of the arms of an interferometer. To eliminate the light transmitted through the interferometer all one needs is to attain a path difference $\lambda/2$ over and interaction length L:

$$\frac{\lambda}{2} = n_2 I_0 S. \qquad (7.21)$$

Switching can also be obtained by using optical cavities. However, one can prove that in cavities the switching action cannot be increased by increasing the interaction length L, as is instead the case for photonic crystals [4].

7.2.5. *Cavities*

Let us consider the system of Figure 7.6. It consists of a layer of thickness a_c and dielectric constant ε_c sandwiched between two periodic

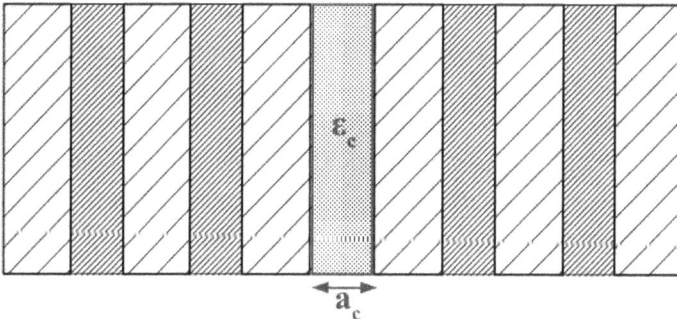

Figure 7.6: Defects can become resonators!

arrays like those of Figure 7.3. The additional layer can be considered as a defect within an otherwise perfect 1D photonic crystal. Each periodic array will behave like a mirror for frequencies within the band gap. Light will be reflected back-and-forth at the edges of the defect layer, and the system can be treated like a Fabry-Perot resonator [5]. The transmittance of such a resonator is given by [3]:

$$T(\omega) = \frac{(1 - R(\omega))^2}{1 + [R(\omega)]^2 - 2R(\omega)\cos[\delta(\omega)]}, \tag{7.22}$$

where

$$\delta(\omega) = 2k_{c,z}(\omega)a_c + 2\varphi(\omega), \tag{7.23}$$

$R(\omega)$ is the (frequency-dependent) reflectivity of each half of the 1D photonic crystal, $\varphi(\omega)$ is the phase shift induced by the reflection by the photonic crystal and

$$k_{c,z} = \frac{\omega}{c}n_c\cos(\theta_c) \tag{7.24}$$

is the component of the wave vector in the z-direction (perpendicular to the stacks).

$T(\omega)$ reaches its maximum (resonance condition) for:

$$\delta(\omega) = 2\pi m, \ m \text{ integer.} \tag{7.25}$$

Based on these considerations, a resonant state can be engineered to appear in the center of a photonic band gap, as shown in Figure 7.7.

Figure 7.8 shows an experimental realization of the microcavity concept [7]. A defect layer containing an organic emitter (F8BT) was sandwiched between two 1D photonic crystals made of alternating layers of two different polymers. The transmittance spectrum of the defect layer alone, without photonic crystals, is shown in Figure 7.8a. A broad minimum around 450 nm is evident, which is due to the dye. The photonic band gap is between about 500 and 600 nm. The defect state shows as a narrow transmittance peak at ∼520 nm. Figure 7.8b) compares the emission spectrum of the dye with that of the photonic crystal. Narrowing of the emission and increased intensity are quite evident. Narrowing of the emission, and intensity increase can be used to fabricate lasers [8], as shown in Figure 7.9.

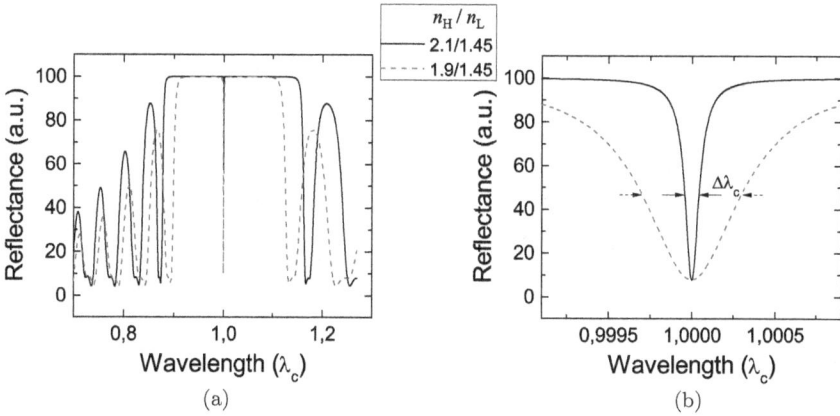

Figure 7.7: Calculated reflectance of cavities composed of two 1D photonic crystals and a cavity layer at normal incidence. The layer design was 10 high-index (H) layers, 10 low-index (L) layers of the same thickness, then H 2L H (L H)10. (a) The resonance appears at the cavity wavelength λ_c, in the center of the stop band. (b) Note the narrow width of the resonance. Adapted with permission from [6]. © Springer.

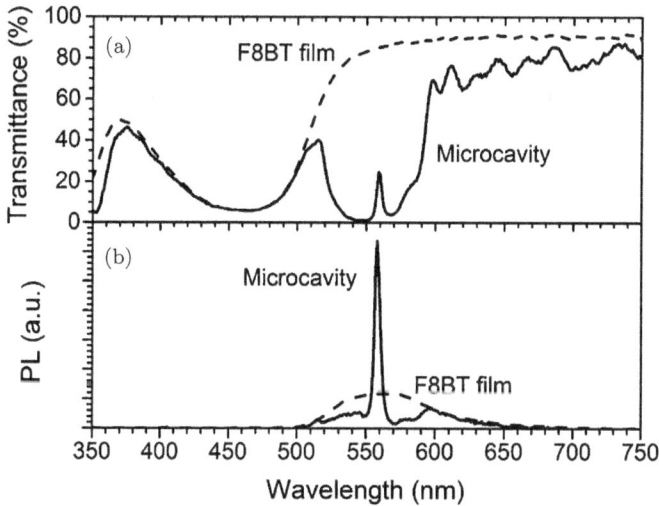

Figure 7.8: (a) Transmittance spectra for both PVK:CA microcavity with an F8BT defect layer and neat F8BT film. (b) Fluorescence spectrum of the PVK:CA microcavity doped with F8BT and of F8BT film. Adapted with permission from [7]. © IOP Publishing.

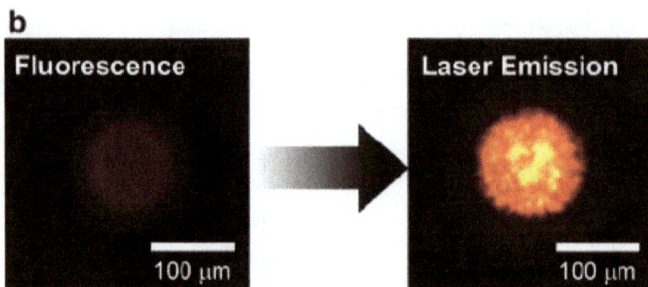

Figure 7.9: Left: fluorescence of a film of the dye rhodamine 640. Right: laser emission from the same film, embedded into a 1D photnic crystal. The excitation power was the same for both experiments. Adapted with permission from [8]. © Wiley-VCH.

7.3. 2D and 3D photonic crystals

The optical properties of 2D- and 3D- photonic crystals can be simulated using the methods reported in Section 1.1. Figure 7.10 reports the band structure of a (rather typical) 2D structure, obtained by drilling holes into a high-dielectric constant material. The simulation refers to propagation into the plane of the crystal. Since the system is invariant under specular reflection, the electromagnetic eigenmodes can be classified as even or odd with respect to this mirror symmetry. Even states are called H- (or TE-) polarized modes, and odd states are called E- (or TM-) polarized. The band structure in Figure 7.10 depends strongly on polarization, yet a photonic band gap (red region) can be achieved. Band structures that include gaps have been calculated for several 3D photonic crystal geometries. However, difficulties in fabricating large-scale, defect-free 3D crystals [9] have so far prevented their widespread use. We will therefore focus on 2D crystals, and more specifically, on defects and their usefulness. This does not reflect poorly, of course, on the marvel of naturally occurring photonic crystals, such as those shown in Figure 7.11 [10].

7.3.1. *Photonic slabs and waveguides*

A most convenient way of controlling and studying light propagation is to integrate a 2D photonic crystal (PC) into a waveguide. The most immediate way of achieving this is to "write" a PC into a slab of a

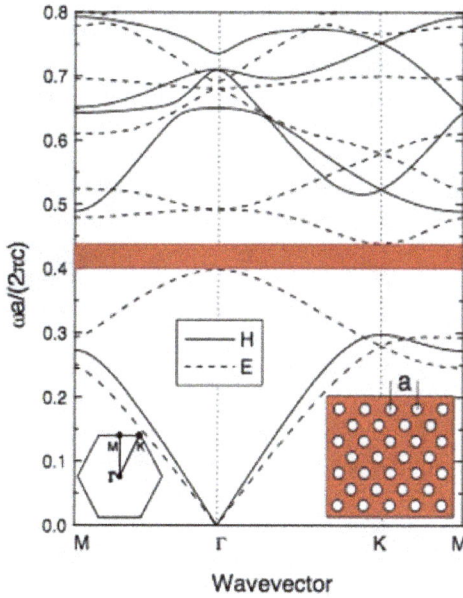

Figure 7.10: Dispersion of photonic bands in a triangular lattice of air holes with radius $r/a = 0.45$ in a dielectric material with $\varepsilon = 12$. Solid (dashed) lines represent H- (E-) polarized modes. The shaded area denotes the photonic gap. Insets: structure and 2D Brillouin zone, respectively. Adapted with permission from [3]. © Springer.

high index material. The slab can be made free-standing in air, or deposited on a substrate with a lower index of refraction. Figure 7.12 shows an example of a free-standing photonic crystal slab [11]. The slab is made of Si and acts as a waveguide. Like any other waveguide, light will be able to propagate within the plane of the slab when the angle of incidence is larger than the critical angle θ_c, as shown in Figure 7.13. The critical angle is given by:

$$\sin(\theta_c) = \frac{n_{clad}}{n_{core}}, \tag{7.26}$$

where n_{core} is the index of the waveguide and n_{clad} the index of the surroundings. Eq. (7.26) translates into the light confinement condition:

$$\omega < \frac{ck_x}{n_{clad}}. \tag{7.27}$$

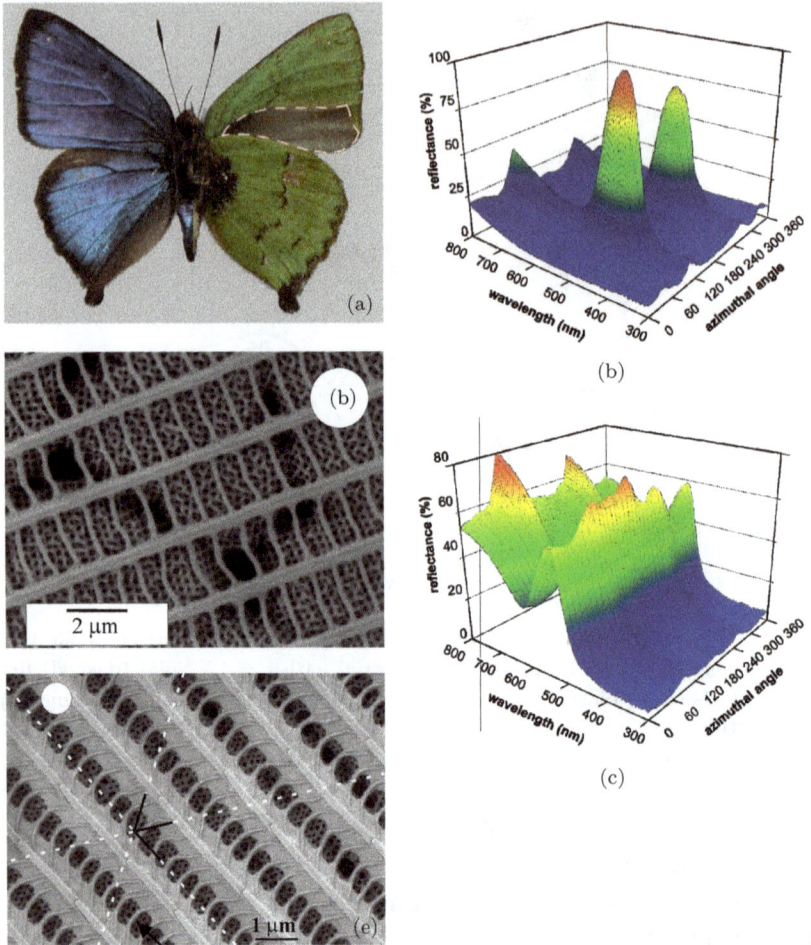

Figure 7.11: An example of naturally occuring 3D photonic crystals. (a) Images of Cyanophrys Remus. Left: dorsal wing (blue). Right: ventral wing (green). (b) Dependence of the dorsal reflectance on wavelength and azimuthal angle. The angle of incidence and collection of the light was 45°. (c) Same, for the ventral wing. (d) Scanning electron micrograph of the ventral wing. (e) Same, dorsal wing. Adapted with permission from [10]. © AK Journals.

Eq. (7.27), denoted as "light line", separates the photonic band into two regions. Modes that fulfill Eq. (7.27) can propagate into the waveguide; all others will escape out of the slab into the surroundings, as shown in Figure 7.14 [3].

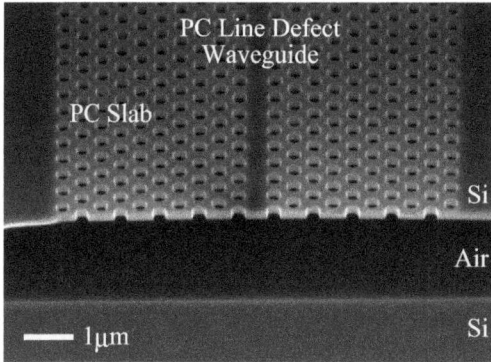

(a)

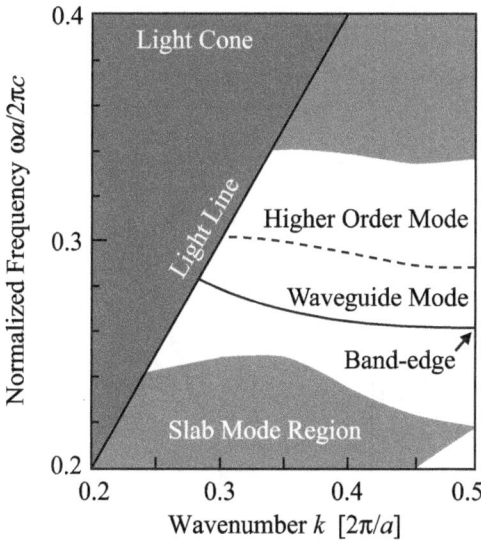

(b)

Figure 7.12: (a) An example of a W1 waveguide in a free-standing Si membrane. (b) Corresponding photonic bands. Note the light line. Adapted with permission from Baba & Mori (2007); © IOP Publishing.

7.3.2. Defects in slabs

Figure 7.12(a) shows a defect in a photonic crystal slab, consisting of a missing row of holes. This defect is called a W1 waveguide. It introduces a defect state within the band gap and in the guided

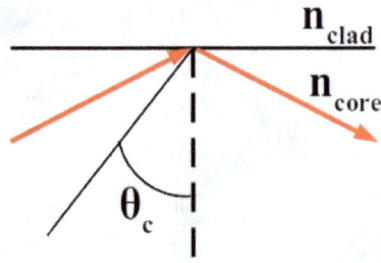

Figure 7.13: Principle of operation of a waveguide: total internal reflection is only achieved for angles of incidence $>\theta_c$.

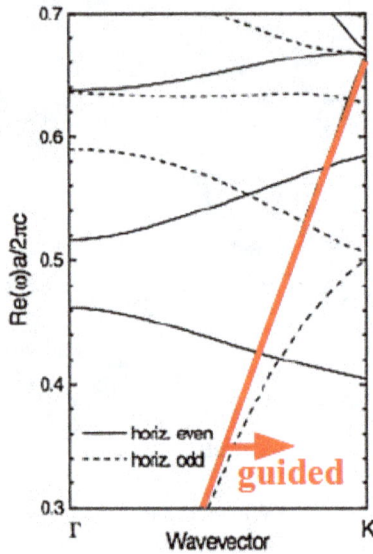

Figure 7.14: Photonic band gap in a waveguide. Red line: light line. Only modes in the hemispace indicated by the arrow will propagate through the waveguide. Adapted with permission from [3]. © Springer.

region, as shown in Figure 7.12(b). The physical meaning of is that light with k_x, ω lying on the defect state and below the light line will be confined within the defect and the slab. The defect will act as a waveguide, allowing light to be bent without the use of mirrors, as shown in Figure 7.15 [11].

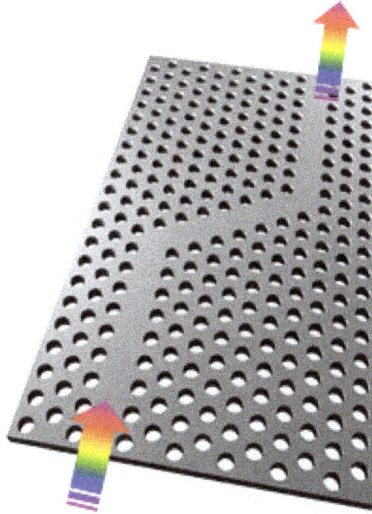

Figure 7.15: Example of light guiding using a defect in a photonic crystal. Adapted with permission from [11]. © World Scientific.

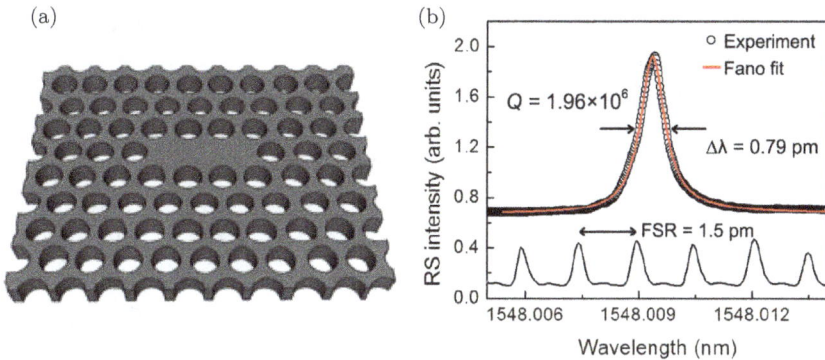

Figure 7.16: (a) Schematic structure of an L3 cavity a photonic crystal slab. (b) Measurement of the Q-factor of a L3 nanocavity in silicon. Adapted with permission from [12]. © Elsevier.

Figure 7.16 shows a point defect, obtained by omitting 3 holes in a 2D photonic crystal. This defect, called L3, can be used as a cavity, much as the defect in 1D PC which was illustrated in Section 1.2.5. Compared to the 1D cavity, the advantage of a L3 defect is that light

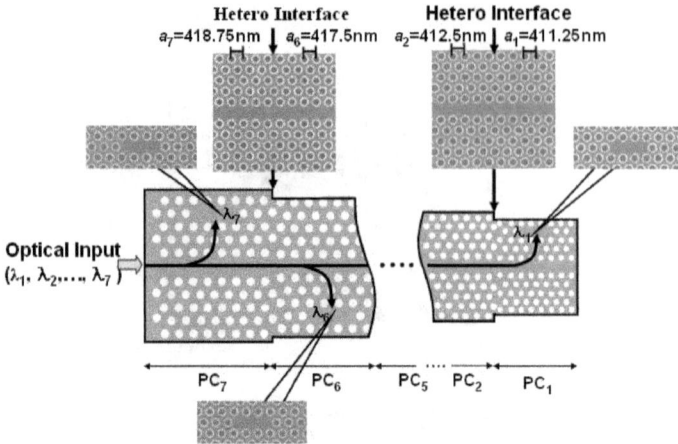

Figure 7.17: Schematic of hetero-PCs consisting of seven PCs, where lattice-constant difference between the neighboring PCs was designed to be 1.25 nm. The insets show the SEM images of representative cavities and hetero-interfaces (the latter are boundaries between different PCs). Adapted with permission from [11]. © World Scientific.

is confined by the waveguide as well as by the cavity. Exceedingly narrow emission lines can be attained using L3 cavities, as shown in Figure 7.16(b) [12].

Figure 7.17 shows a prototype optical circuit. Light of 6 different wavelengths is injected into a W1 waveguide. The spacing of the photonic crystal surrounding the waveguide is changed along its length to reject one of the wavelengths each time a PC boundary is crossed. A L3 defect is placed next to the waveguide and is designed to act as a cavity for the rejected wavelength. The resulting circuit is capable of lasing at multiple wavelengths, as shown in Figure 7.18 [12].

7.4. Fabrication techniques

In the following, some popular techniques used for the fabrication of photonic crystals will be described. Given the creativity of researchers working in this area, the list is not exhaustive. Rather, it is meant as a set of general fabrication guidelines.

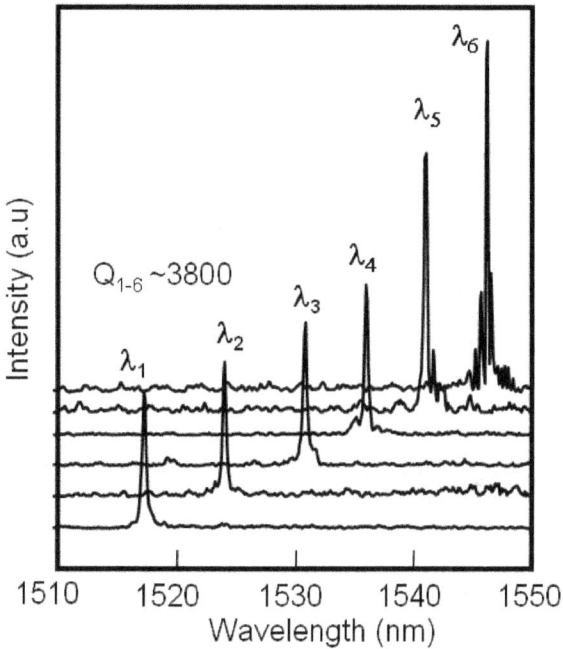

Figure 7.18: Emission of the cavities of Figure 7.17. Adapted with permission from [11]. © World Scientific.

7.4.1. *Microfabrication techniques*

One-dimensional photonic crystals can be conveniently fabricated using standard microfabrication techniques such as spin- and sputter-coating. Using these techniques, thin films of materials with different refraction index and thickness can be stacked upon each other. Phosphors can be easily incorporated into defects [13]. 2D photonic slabs and waveguides can also be fabricated using standard techniques. One intuitive technique is focused ion beam (FIB). In this technique, an ion beam is used to drill holes into a substrate. FIB, however, has a low throughput. Electron beam- or photo-lithography, coupled with reactive ion etching, are encountered more often. The reason is that these are higher-throughput techniques, which are also compatible with standard protocols of the semiconductor industry.

7.4.2. Holographic techniques

A set of powerful techniques is based on interference of coherent beams. Interference generates periodic patterns, which are of course ideal for photonic crystals.

The most intuitive holographic set-up is shown in Figure 7.19. A laser beam is sent through a beam splitter and from this onto two mirrors. The mirrors reflect the light onto a substrate coated with a photosensitive polymer. Interference fringes form on the substrate with a spacing d given by:

$$d = \frac{\lambda}{2\sin(\frac{\theta}{2})}, \qquad (7.28)$$

where λ is the wavelength of the incident light and θ the incident angle, shown in Figure 7.19. Development (i.e., removal) of the unexposed part of the coating yields a 1D periodic pattern. While simple and intuitive, the set-up of Figure 7.19 yields only 1D photonic crystals. It is also quite sensitive to vibrations: four components (the beam splitter, the two mirrors and the substrate) must remain in

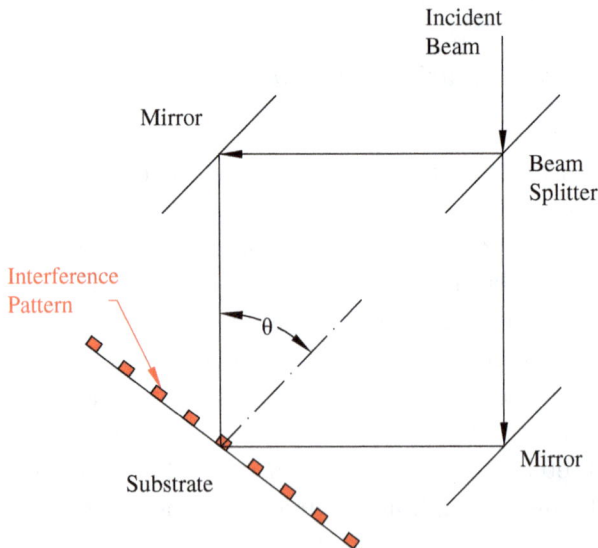

Figure 7.19:   Schematic representation of an interference lithography set-up.

the same position during exposure, otherwise the interference pattern will be washed out.

A more practical approach, which also yields 2D PCs, is shown in Figure 7.20. In this approach, the beam is expanded to illuminate three diffraction gratings, etched on the same substrate. This set-up reduces the number of optical components in the beam path, and therefore alleviates vibrational issues. Interference of the beams yields a set of pillars, shown in Figure 7.21 [14]. This approach can be extended to the fabrication of 3D photonic crystals. A set-up is shown in Figure 7.22 [15]. A beam is expanded to illuminate a prism with a triangular base. Each side face of the prism refracts the beam towards the center. A fourth beam, coming from the base at the top, provides a depth modulation. This modulation creates a three-dimensional interference pattern, shown in Figure 7.22(d). Since the

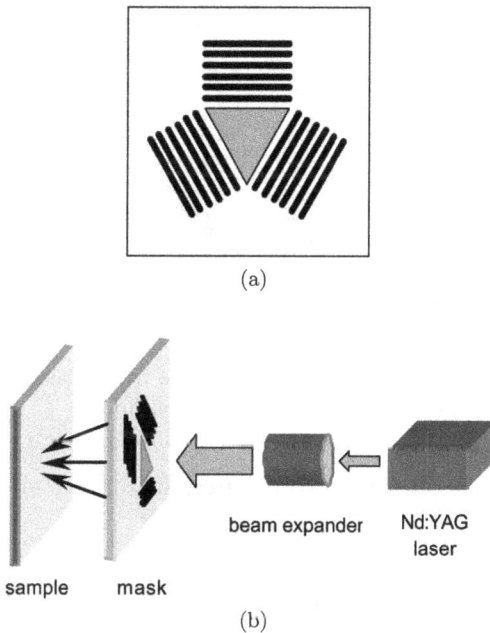

(a)

beam expander   Nd:YAG laser

sample   mask

(b)

Figure 7.20: Schematic representation of a set-up tused to fabricate 2D photonic crystals using a reduced set of optical components. Adapted with permission from [14]. © AIP Publishing.

(a)

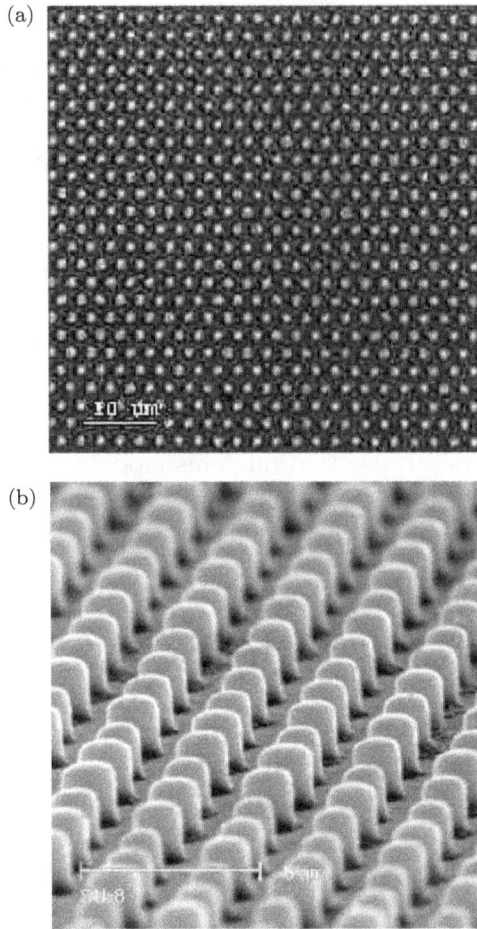

(b)

Figure 7.21:   Pillars formed by the interference of the beams of Figure [7.20].
Adapted with permission from [14]. © AIP Publishing.

prism can be laid on top of the coated substrate, vibrational issues
are alleviated.

7.4.3. *Pulsed laser techniques*

In these techniques, a laser is employed to drill a hole or densify
the exposed region. In almost all cases a pulsed laser is employed.
The reason is the high power that short pulses can deliver to the

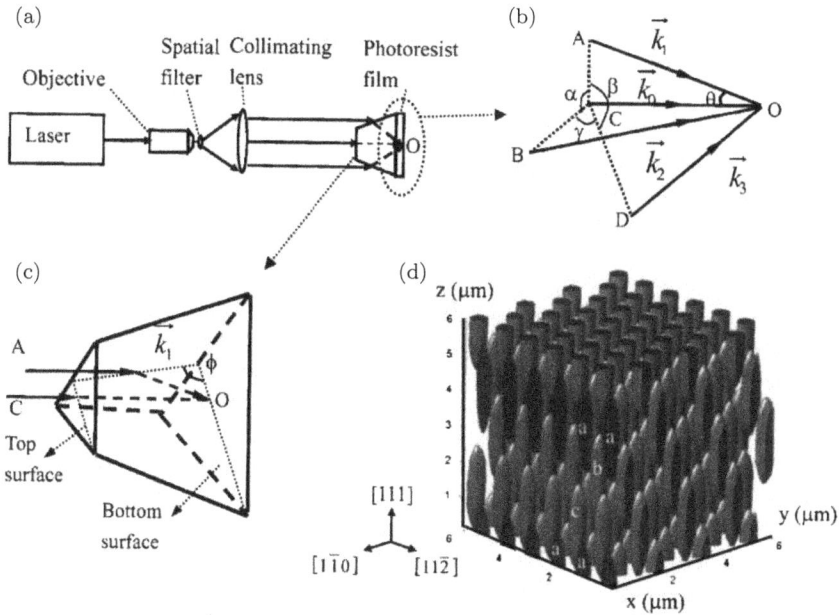

Figure 7.22: Schematic of a holographic technique used for the fabrication of 3D photonic crystals. (a) Schematic set up. (b) Beam angles and paths. (c) prism used to produce the beams. (d) calculated 3D interference pattern. Adapted with permission from [15]. © AIP Publishing.

target. For example, a typical pulsed laser (Yittrium-Aluminum-Garnet, YAG) will deliver pulses with a typical energy of 10 mJ and a typical duration of 5 ns. The power of such a pulse is

$$P = \frac{E}{t} = \frac{10 \times 10^{-3} J}{5 \times 10^{-9} s} = 2\text{MW}, \qquad (7.29)$$

which is sufficient to melt glass (or vaporize metals, for that matter).

A most immediate application of laser densification is represented by fiber gratings: an interference pattern is created perpendicular to the axis of an optical fiber. The index is changed in the exposed regions, giving rise to a 1D photonic crystal. This crystal will reject selected wavelengths. Since the wavelengths depend on the lattice spacing, applications of fiber gratings as stress and temperature sensors are immediate.

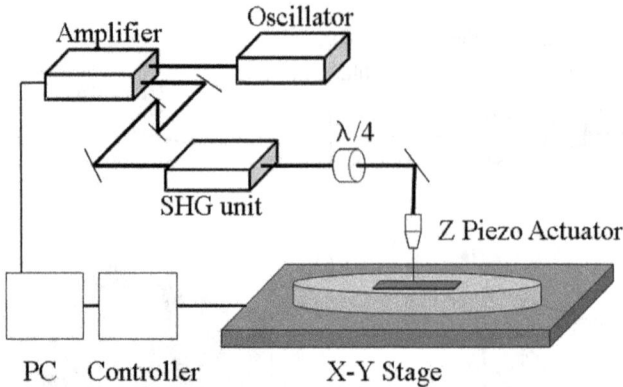

Figure 7.23: Schematic representation of a direct write laser set up used for the fabrication of 2D photonic crystals. Adapted with permission from [16]. © OSA Publishing.

To fabricate 2D photonic crystals, the substrate is scanned under a focused laser, as shown in Figure 7.23 [16]. In the specific example, the substrate was glass doped with Zn and S. Laser heating led to formation of ZnS nanoparticles in the exposed regions. The woodpile structure produced by the set-up is shown in Figure 7.24 together with transmission spectra for different values of the spacing between lines. A variation of this technique, where the laser is focused at different depths within the glass substrate, can be used to densify selected spots in 3D. The resulting structures resemble (but luckily are not) trinkets sold at tourist traps like Venice or Times Square, and an example is shown in Figure 7.25 [17].

7.4.4. *Opal fabrication*

A very popular 3D photonic crystal is the (inverse) opal, shown in Figure 7.26 [18]. In principle, this structure is not difficult to fabricate. Dispersions of monodisperse nanoparticles can be prepared by following one of the many recipes reported in the literature. For example, the Stöber method [19] can be used to synthesize SiO_2 spheres. The dispersion is then concentrated to a syrup, and the opal is formed by precipitation or solvent evaporation. The evaporation step is also the crux of the problem,

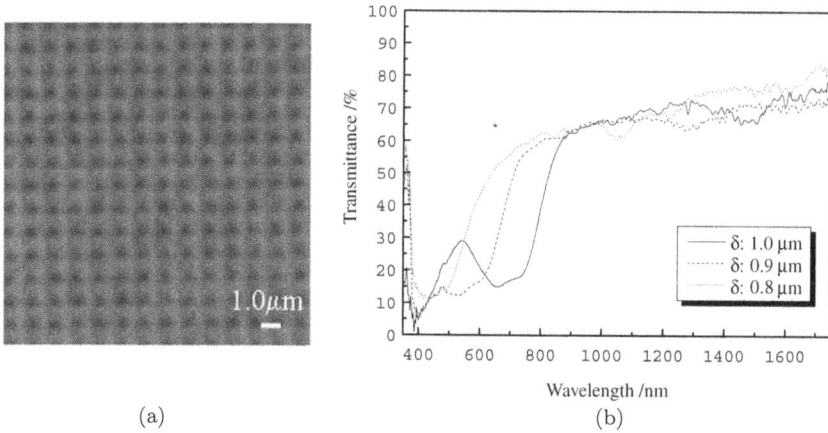

1.0μm

(a)

(b)

Figure 7.24: Left: Woodpile structure created using the set-up of Figure 7.23. Right: transmittance spectra for different values of the spacing between lines. Adapted with permission from [16]. © OSA Publishing.

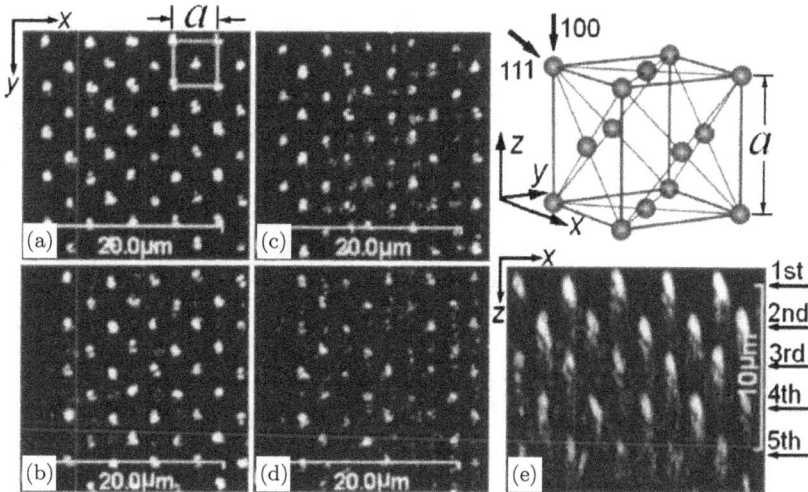

Figure 7.25: Confocal microscope images of the four top layers (a–d) of a 3D photonic crystal fabricated by laser densification of selected regions of a glass substrate. (e) x-z image, showing the top five layers of the structure. The unit cell is shown in the upper right corner. Adapted with permission from [17]. © AIP Publishing.

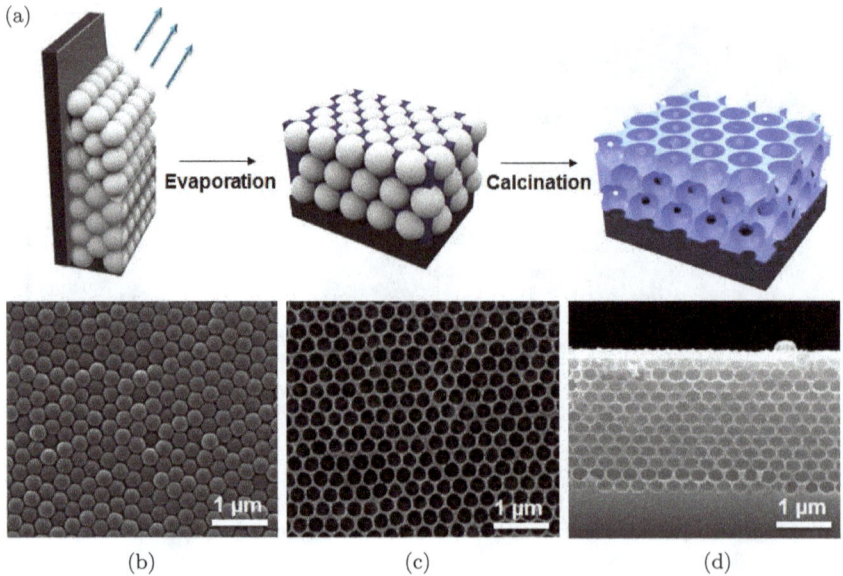

Figure 7.26: Fabrication of opals and inverse opals. (a) A concentrated suspension of polymer nanoparticles is evaporated from a vertical substrate. After evaporation, the opal is infiltrated with an oxide-forming organometallic compound. Calcination removes the polymer nanoparticles and leaves an inverse opal behind. (b) Image of the opal. (c, d) Top and side views of the inverse opal. From [18]. © MDPI.

since we need the spheres to form an ordered lattice. The formation of an ordered lattice is time-consuming and hardly upscalable; this has limited the widespread application of artificial opals. A proven assembly method is vertical deposition, shown in Figure 7.26. In vertical deposition, the solution climbs on the substrate walls due to capillarity, and results in high quality opals. To fabricate an inverse opal, say, of an oxide material, a polymeric opal is infiltrated with a solution of an alkoxide such as tetramethyl orthosilicate. Hydrolysis-condensation of the alkoxide is catalyzed by addition of a dilute base such as NH_4OH. After SiO_2 has formed in the voids, the polymeric component is removed by calcination, as shown in Figure 7.26. To form a polymeric inverse opal, one starts from a SiO_2 opal. The opal is then infiltrated with a solution of a monomer and an initiator. After polymerization, the SiO_2 beads are dissolved using concentrated NaOH (preferable to the more popular, but hazardous, HF dissolution route).

References

1. K. Sakoda, "Optical Properties of Photonic Crystals", Springer, Berlin (2004).
2. A. Yariv, P. Yeh, "Photonics: Optical Electronics in Modern Communications", 6th edn. Oxford Series in Electrical and Computer Engineering (Oxford University Press, Oxford, 2006)
3. M. Liscidini and L. C. Amerini, "Photonic Crysals: And introductory survey", in D. Comoretto, Ed., "Organic and hybrid photonic crystals", Springer, Heidelberg, 2015.
4. T. F. Krauss, *J. Phys. D: Appl. Phys.* **40**, 2666–2670 (2007).
5. E. Hecht, "Optics", 5th Edition, Addison-Wesley (2015).
6. R. Brückner, V. G. Lyssenko, and K. Leo, "Plasmonic and Photonic Crystals", in D. Comoretto, Ed., "Organic and hybrid photonic crystals", Springer, Heidelberg, 2015.
7. G. Canazza, F. Scotognella, G. Lanzani, S. de Silvestri, M. Zavelani-Rossi, D. Comoretto, *Laser Phys. Lett.* **11**, 035804 (2014).
8. S. Furumi, J. Polym, **579** (2013). S. Furumi, H. Fudouzi, H.T. Miyazaki *et al.*, *Adv. Mater.* **19**, 2067 (2007).
9. S. Takahashi, K. Suzuki, M. Okano, M. Imada, T. Nakamori, Y. Ota, K. Ishizaki, S. Noda, *Nat. Mater.* **8**, 721–725 (2009).
10. L. P. Biró, Z. Bálint, Z. Vértesy, K. Kertész, G. I. Márk, V. Lousse and J. Vigneron, *Nanopages* 1 (2006) 2, 195–208.
11. B. Song, T. Asano and S. Noda, *NANO: Brief Reports and Reviews* Vol. 2, No. 1 (2007) 1–13.
12. Y. Lai, S. Pirotta, G. Urbinati, D. Gerace, M. Minkov, V. Savona, A. Badolato, M. Galli, *Appl. Phys. Lett.* **104**, 241101 (2014).
13. R. M. Almeida, M. C. Goncalves, S. Portal, *Journal of Non-Crystalline Solids* 345&346 (2004) 562–569.
14. T. Meyer, *Applied Physics Letters,* 2001.
15. L. Wu, Y. Zhong, C. T. Chan and K. S. Wong, *Applied Physics Letters* **86**, 241102 (2005).
16. N. Takeshima, Y. Narita, T. Nagata, S. Tanaka, K. Hirao, *Optics Letters* **30**, 537 (2005).
17. G. Zhou, M. J. Ventura, M. R. Vanner and M. Gu, *Applied Physics Letters* **86**, 011108 (2005).
18. W. S. Lee, T. Kang, S. H. Kim and J. Jeong, *Sensors* **18**, 307 (2018).
19. W. Stöber, A. Fink, E. Bohn, (January 1968). *Journal of Colloid and Interface Science.* **26**(1): 62–69.

Chapter 8

Electrically Conducting Polymers and Their Applications

8.1. Electrically conducting polymers

To understand how one can introduce electrical conductivity into a polymer, let us look at the molecular structure of the simplest polymer of this class, polyacetylene.

Polyacetylene, shown in Figure 8.1, is characterized by an alternation of single (σ) and double (σ, π) bonds. The side view of the polymer chain, shown in Figure 8.2, is most illuminating. Each atom pair shares a π orbital with one of its nearest neighbors. Now, let us assume to remove one electron from a π bond (red line in Figure 8.1). The hole created allows for an electron of the neighboring atom to "jump ship", as shown in Figure 8.1. Electrons can move along the polymer backbone following the same scheme.

So, how is removal of an electron achieved? The simplest way is to contact the polymer chain with a compound that will withdraw electrons from the polymer chain. Donation is also possible, but here we will focus on withdrawal. The choice of compound is determined by the electrochemical reduction potentials, some of which are reported in Table 8.1. The meaning of the reduction potential is as follows: the more negative the potential, the more strongly the equilibrium is moved to the left side of the reaction. So, if we contact polyacetylene (neutral, $[CH]_x$) with Na metal, electrons will move from Na to the polymer chain. Na will be oxidized and the polymer chain will be reduced. Conversely, if we contact the polymer chain with Ag^+,

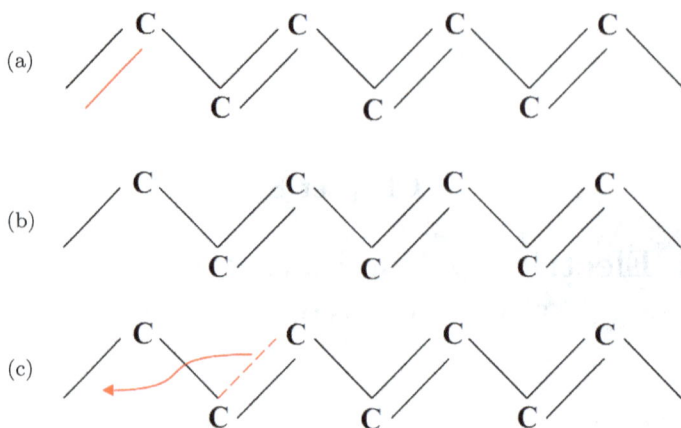

Figure 8.1: (a) Schematic representation of pristine polyacetylene. (b) The electron marked in red is removed from the chain. (c) An electron migrates from the neighboring bond to the now empty orbital.

Figure 8.2: Bonds in polyacetylene.

Table 8.1: Selected reduction potentials of polyacetylene and common dopants.

Couple	$E_{red}(H^+/H_2)$, V
$Li^+ + e^- \rightleftarrows Li$	-3.05
$Na^+ + e^- \rightleftarrows Na$	-2.71
$[CH^{-(q+a)}]_x + (ax)e^- \rightleftarrows [CH^{-q}]_x$	-1.24^*
$[CH^{+a}] + (ax)e^- \rightleftarrows [CH]_x$	-0.57^*
$I_2 + 2e^- \rightleftarrows 2I^-$	$+0.54$
$Ag^+ + e^- \rightleftarrows Ag$	$+0.80$
$(ClO_4)^- + 8H^+ + 8e^- \rightleftarrows Cl^- + 4H_2O$	$+1.37$

* = data from Ref. [1].

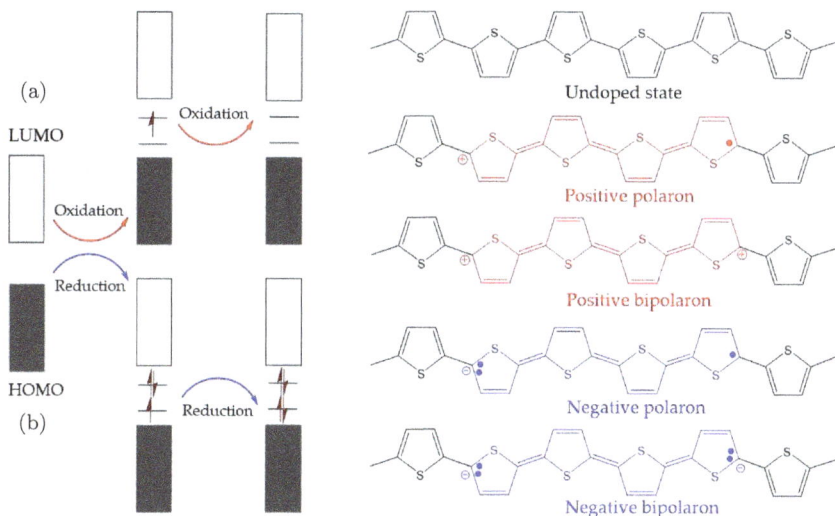

Figure 8.3: Band and chemical structures of polythiophene with (a) p-type doping and (b) n-type doping. Adapted with permission from [2]. © MDPI.

electrons will be transferred from the chain to form metallic Ag. In both cases, introduction or removal of electrons will allow electron transfer between adjacent carbon atoms.

Figure 8.3 shows the structure and electronic levels of polythiophene, a very popular conducting polymer [2]. One should note how doping introduces levels within the HOMO-LUMO gap. One should also bear in mind that charging and charge movements are accompanied by deformation of the polymer chain. The quasiparticle associated with charge transfer and chain deformation is called a polaron. Polarons introduce states in the HOMO-LUMO gap which can be identified spectroscopically, as shown in Figure 8.4 [3]. Also of notice is the evolution from polaron to bipolaron and to bipolaron bands with increasing doping levels, as well as the increase in the HOMO-LUMO gap at the highest doping levels.

Table 8.2 reports the electrical conductivities of commonly employed conducting polymers. One should note that even the best conducting polymer, polyacetylene, has a conductivity which is orders of magnitude lower than copper. Therefore, conducting polymers should not be considered to be the equivalent of metals.

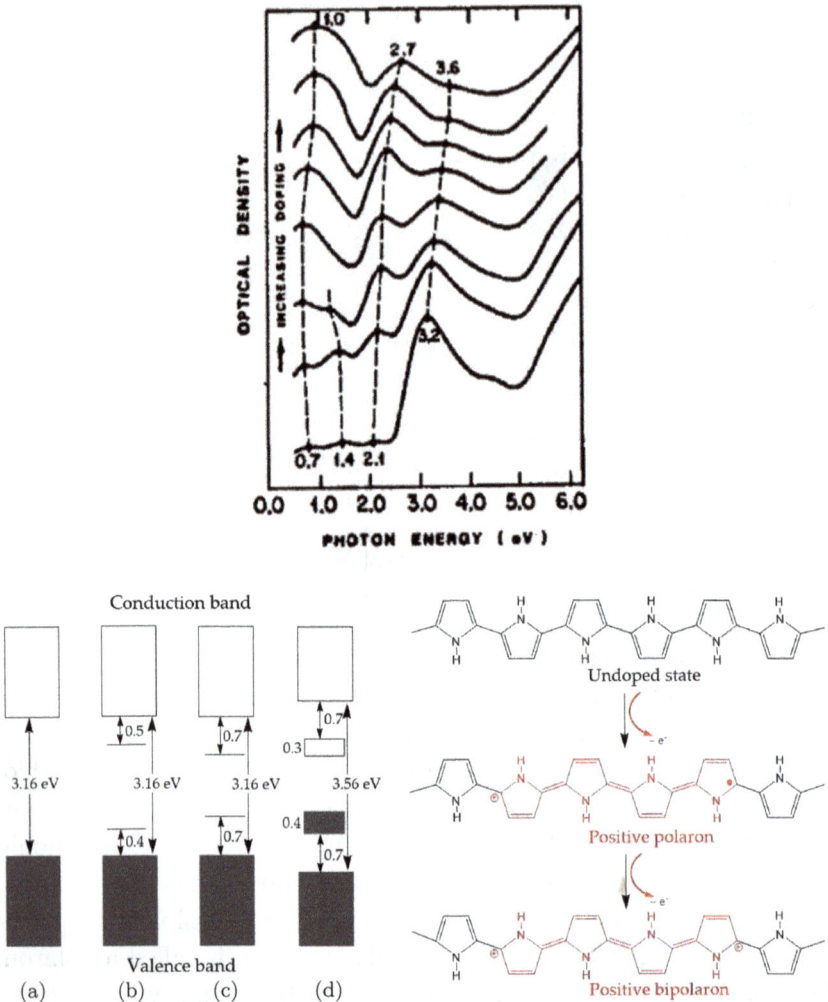

Figure 8.4: Top: Absorption spectra of polyaniline at increasing doping levels. Adapted with permission from [2]. © MDPI. Bottom: corresponding energetic levels. (a) undoped polyacetylene. (b) polaron. (c) bipolaron. (d) polaron bands. Adapted with permission from [3]. © Springer.

One of the reasons of the low conductivity of conducting polymers is disorder. Conducting polymers, like most polymers, are amorphous and highly disordered. In addition, doping is seldom homogeneous, as shown in the drawing in Figure 8.5 [4]. Because of the disorder, charges must hop between adjacent polymer strands. Several models

Table 8.2: Comparison between electrical conductivities of metals and selected conducting polymers.

Material	σ(S/cm)	Polymer
Ag, Cu	10^6	
Fe	10^5	
Bi	10^4	Polyacetylene
	10^3	PEDOT
	10^2	Polypyrrole
	10^1	Polyaniline
Glass	10^{-10}	

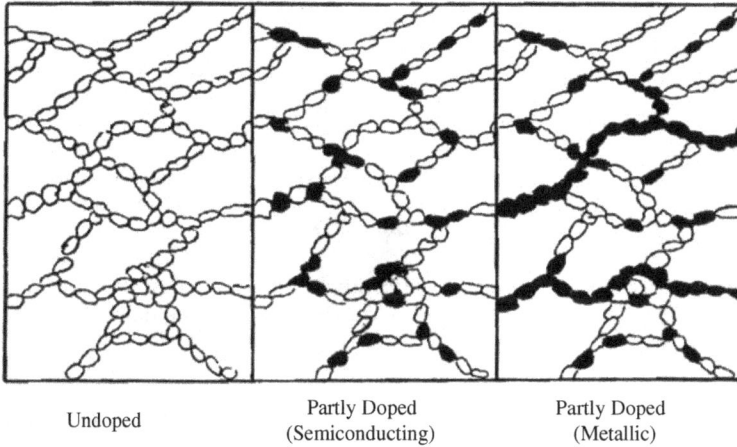

Undoped   Partly Doped   Partly Doped
              (Semiconducting)   (Metallic)

Figure 8.5: Cartoon illustrating formation of percolation pathways for increasing doping levels. Adapted with permission from [4]. © Taylor & Francis.

have been developed to account for the (activated) charge transfer in conducting polymers. A model that works well for polyacetylene is the Mott variable range hopping model [5]. In this model, the conductivity σ is given by:

$$\sigma = \sigma_0 e^{-(\frac{T}{T_0})^{-(\frac{1}{n+1})}},$$ (8.1)

where n is the dimensionality of the system. A fit of Eq. (8.1) to experimental data [6] is shown in Figure 8.6. The fit yields a $T^{-1/4}$

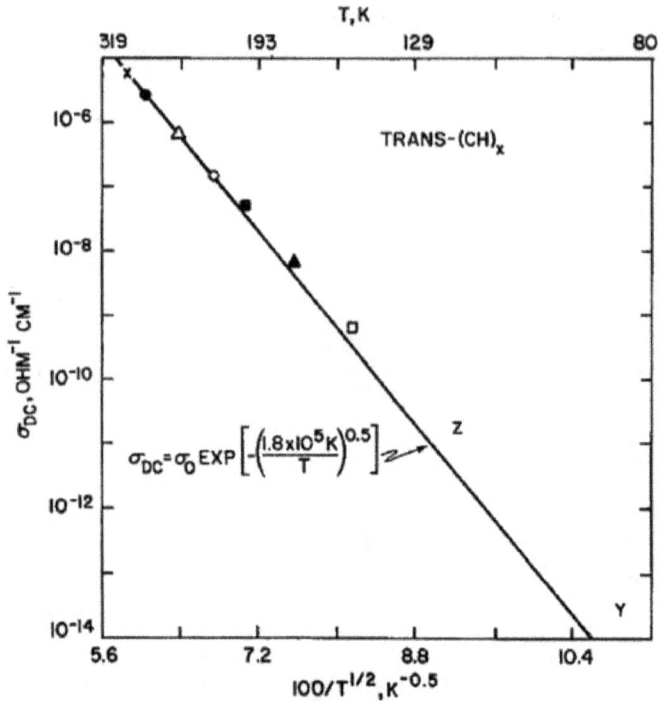

Figure 8.6: Dependence of electrical conductivity of polyacetylene on temperature. The line is a fit of the Mott model to the experimental data. Adapted with
permission from [6]. © Taylor & Francis.

dependence, indicating three-dimensional conduction. One has to
note, however, that conductivity is strongly dependent on disorder
and doping level. Not surprisingly, results and theoretical interpre-
tations reported in the literature are not unanimous.

Selected applications of conducting polymers will now be discussed. Some of these applications are not strictly nano-related. However, they often are good examples of general principles which can
be used for nano-applications.

8.2. Electrochromism

Figure 8.4 is an excellent example of the changes in the optical
properties of conducting polymers (CP) induced by doping. At low
doping levels, the material is nearly transparent in the near IR.

An intense absorption band is evident around 3.0 eV. This band originates from the π–π^* transition and accounts for the green-yellow color of polyaniline at low doping levels. When the doping increases, states appear within the band gap. These states are typically shallow and are responsible for the increase of the optical absorption in the red and near infrared regions of the spectrum. Correspondingly, highly doped CPs are grey or black. Dopants (and therefore the optical absorption) can be inserted and removed by placing a CP in an electrochemical cell; hence the effect is termed electrochromism. Electrochromism of CPs has been known for a long time. However, surprisingly few devices have been commercialized which exploit the electrochromic properties of CPs. There are several reasons for this dearth of practical applications. An analysis of these reasons is useful, since many of the same issues affect nano-devices.

8.2.1. Cyclability

To be useful in electrochromic applications, a CP must be able to be cycled several times (hundreds at a minimum). An example is shown in Figure 8.7, which shows the extent to which commercial devices must be tested in order to be acceptable for market deployment. One important aspect to keep in mind is that cyclability must be demonstrated for all relevant aspects of a device. For example, the switching time must not change. If changes are detected, they must be quantified and documented. Similarly, switching symmetry, charge capacity and dynamic range should be measured during each cycle, and their changes documented.

8.2.2. Durability of the envelope

Figure 8.8 shows an example of a commercial, CP-based electrochromic device, used for sunglasses commercialized by Ashwin-Ushas Corp. The device is flexible, completely transparent in its undoped state, and capable of achieving high contrast (i.e., very dark in its doped state). A schematic representation of the device is shown at the top of Figure 8.8. Two electrodes are fabricated by depositing Indium Tin Oxide (ITO, a transparent, electrically conducting oxide) on thin plastic (Mylar) sheets. Conducting polymers are deposited

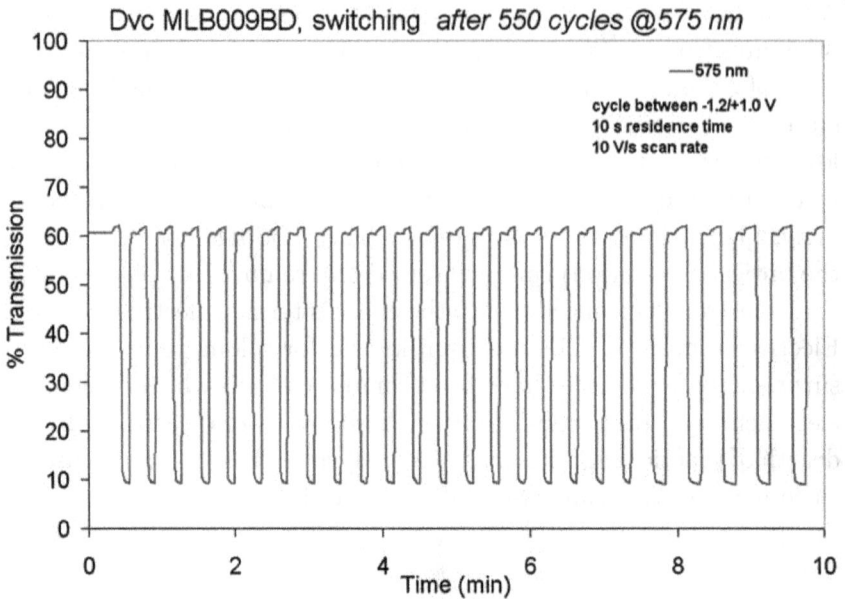

Figure 8.7: Changes in absorption at 550 nm of a conducting polymer used for commercial electrochromic applications. Adapted with permission from [3]. © Springer.

on both electrodes and the space between the electrodes is filled with an electrolyte. The sides of the device are Mylar shims, glued to the electrodes. Such a device has to be able be able to withstand multiple cycles of electrochromic switching, mechanical solicitations (e.g., bending), of temperature changes, humidity and solar irradiation. These tests are time-consuming and not particularly exciting for a basic scientist, but all-important for commercial applications.

8.3. Sensors

The strong dependence of the electrical conductivity on doping makes CPs candidates for sensing applications. One such sensor is described in Refs. [7–9] and references therein. Initial work showed that exposure to ultraviolet radiation during polymerization led to formation of polyaniline (PANI) nanofibers, as shown in Figure 8.9. The high surface area of the nanofibers increases the response and decreases

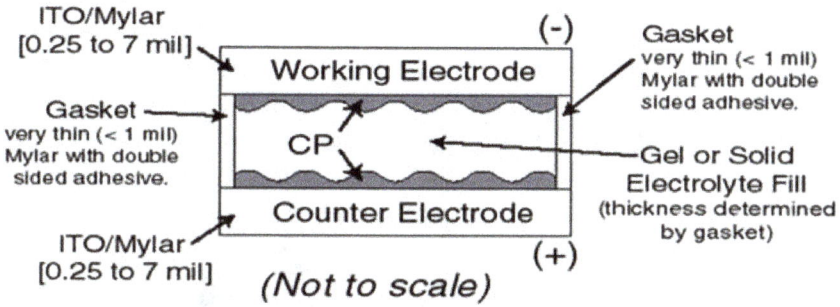

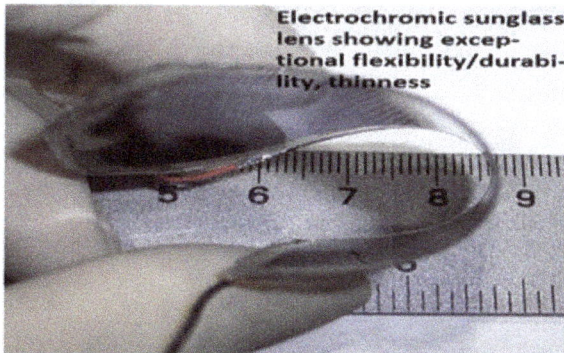

Figure 8.8: Top: Schematic and actual image of commercial electrochromic sunglass lens. Bottom: prototype showing high light/dark contrast. Adapted with permission from [3]. © Springer.

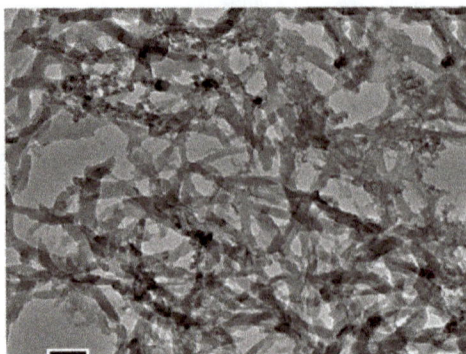

Figure 8.9: Bright field transmisstion electron micrograph of polyaniline nanofibers. The darker spots are Ag nanoparticles. The scale bar is 100 nm. Adapted with permission from [9]. © Elsevier.

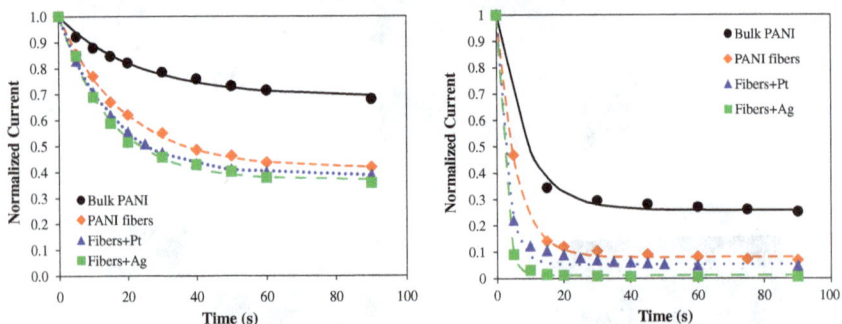

Figure 8.10: Response of polyaniline sensors to (a) Toluene and (b) triethylamine vapor. Adapted with permission from [9]. © Elsevier.

the response time, making nanofibers a promising candidate for chemosensors. The nanofibers can be deposited across electrodes and decorated with noble metal nanoparticles (Ag, Pt, Au), with minimal processing. Figure 8.10 shows the response of such a sensor to toluene and triethylamine vapor, respectively. For both vapors, one notices a decrease of the current flowing through the sensor (i.e., an increase in resistivity, since the measurements are taken at a constant applied voltage). The current decrease is larger for fibers than for bulk polyaniline, as expected from surface area considerations. For the triethylamine system, current decreases by more than one order of magnitude, due to the reaction of the analyte with the dopant. Addition of metal nanoparticles increases the sensitivity, likely due to the strong interactions of metals with the analyte. This strong interaction increases the sticking coefficient of the analyte, which results in greater resistivity changes. For toluene, the decrease in current is only of about 2 times and it is not strongly affected by the presence of metal nanoparticles. This is because toluene does not react with the dopant, or the metal nanoparticles. Toluene induces, instead, polymer swelling. Swelling separates the fibers and decreases the current. Thus, two different gases lead to the same effect (increase in resistivity), but for entirely different reasons. Lack of a specific response is a relevant issue for sensing applications, which often prevents even very promising systems from reaching the application stage.

8.4. Light Emitting Devices (LED)

These most important devices, and their close relatives, organic and dye-sensitized solar cells, are at the forefront of research efforts. The working principle of CP-based LEDs is shown in Figure 8.11, together with a cross-section of a device [10]. To understand how the device works, let us start from its cross-section. A glass (or plastic) substrate is coated with ITO, which represents the anode of the circuit. Two layers of different CPs are then deposited on top of the ITO. The lower layer, in contact with the ITO, is a hole-conducting CP. The upper layer is an electron-conducting CP. This CP has also a high luminescence quantum efficiency and is the light-emitting region of the device. The topmost layer is a metal, whose function is to inject electrons into the system. The band diagram of Figure 8.11 is key to understand the flow of charge through the device. Let us start from the metal contact. Ideally, the Fermi energy of the metal should lie above the energy of the LUMO of the CP. This is the case for Ca, whose use would lead to a spontaneous injection of electrons into the device. Alas, Ca is not a practical choice. One has to resort to Au, or Al, which introduce an activation barrier for electron transfer. Fortunately, this barrier can be overcome by applying a comparatively small voltage. Now, let us move to the ITO side. Here, we have to

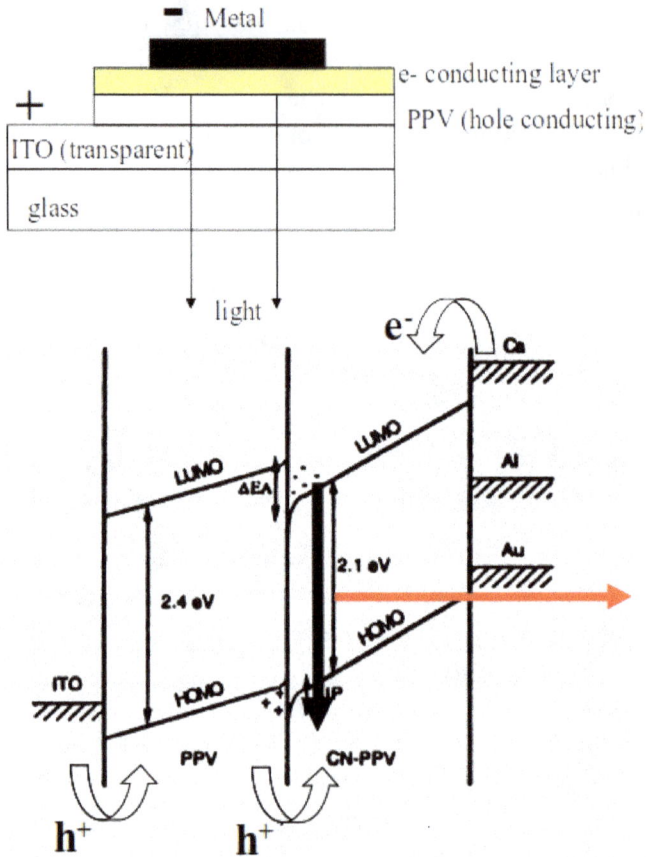

Figure 8.11:   Top: cross-section of a light emitting diode based on conducting polymers. Bottom: energetic levels and direction of motion of holes and electrons. Adapted with permission from [10]. © Materials Research Society.

inject holes into the system, which is equivalent to say that electrons must flow from the CP to the ITO. Ideally, the Fermi level of the ITO should lie below the level of the HOMO of the CP. When this is not the case, a barrier ensues, which requires an applied voltage to be overcome. Now, let us look at the junction. The levels of the HOMO and LUMO of the two CPs must be as shown in Figure 8.11. That is, the LUMO of the electron-conducting CP must be lower than the LUMO of the hole-conducting CP. Vice versa for the HOMOs. In this way, electrons injected by the metal will accumulate at the

(a)

(b)

(c)

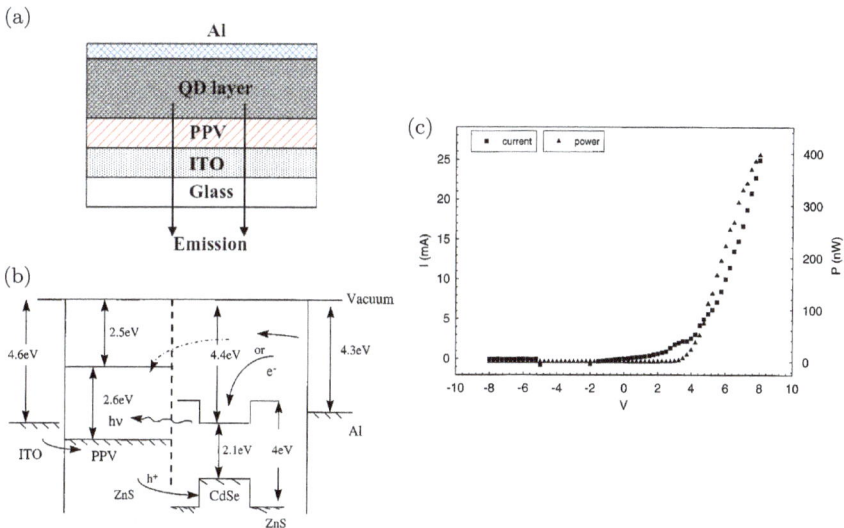

Figure 8.12: (a) Cross-section od a CP-quantum dot LED. (b) Energy levels. (c) I-V and light emission-V curves. Adapted with permission from [11]. © American Physical Society.

junction and will not be able to cross into the hole-conducting side. The same applies to the holes on the other side of the junction. The final condition is that the barrier at the LUMO side, ΔE_A, must be larger than the barrier at the HOMO side, ΔI_P. In this way, electrons will be able to jump from the HOMO of the electron-conducting CP to the HOMO of the hole-conducting CP. Put in another way, the holes will have an easier job to cross the junction than electrons. Now for the kick: when we say that a hole crosses the junction, what really happens is that an electron is removed from the HOMO of the electron-conducting CP. An electron in the HOMO, which has nowhere to go because of the large barrier ΔE_A, will recombine with the hole and emit light in the process.

Introduction of quantum dots in the system adds tunability of the emission wavelength. A schematic of a quantum dot-based device is shown in Figure 8.12(a), together with the corresponding energy diagram [11]. The energy of the conduction band of the chromophore (CdSe) increases with decreasing particle size; hence there is no risk of particle size reduction interfering with the energy diagram. I-V

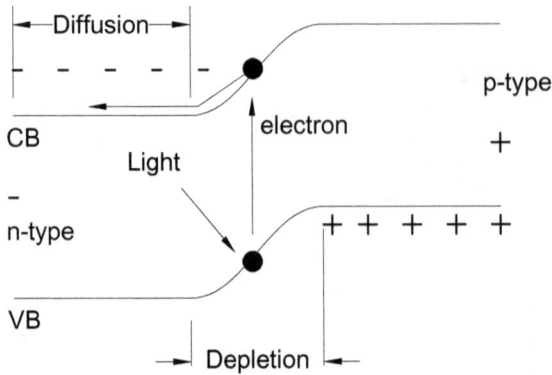

Figure 8.13:   Schematic representation of a p-n junction used as a photovoltaic cell.

and light intensity-V curves are reported in Figure 8.12(c). Light emission and current grow at the same rate with applied voltage, confirming that the device does indeed behave as a LED.

8.5. Photovoltaics

A good starting point for the understanding of the role of nanotechnology in solar cells is the conventional, bulk silicon solar cell.

In a photovoltaic cell, a p-type semiconductor and an n-type are brought together. Electrons flow spontaneously from the n- to the p-side of the junction. This charge flow creates an electric field which opposes the motion of the charges. At equilibrium, the situation looks like the one in Figure 8.13 and the I–V curve looks like the black trace in Figure 8.14. If now we shine light on the device, an electron will be promoted from the VB to the CB. Assume to short-circuit the p-n junction. The electron will flow towards the n- side of the junction. However, since the voltage under short circuit conditions is zero (or negligible), the electron will have zero (or negligible) energy. Assume now to apply a forward bias, as shown in Figure 8.14. A rather bizarre situation occurs. The photogenerated electrons will flow towards the n-type region, while some electrons will flow from the n-type region to the p-type region. These latter electrons are those that would flow

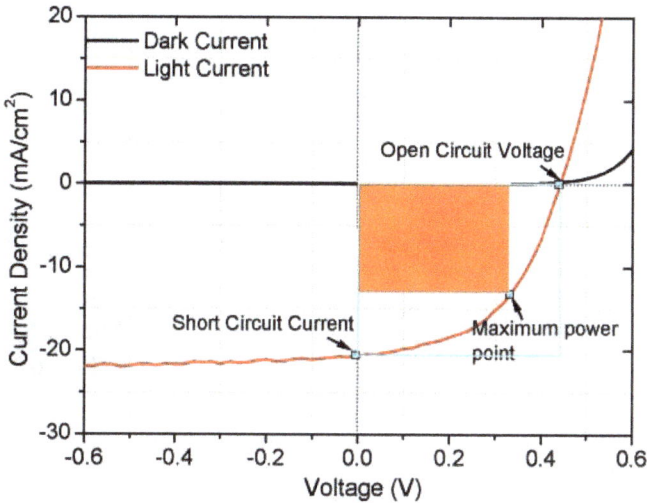

Figure 8.14: Black trace: I–V curve of a p-n photovoltaic cell without illumination. Red trace: same, under illumination. Adapted with permission from [12]. © Wiley-VCH.

through the circuit in forward bias. Put these concepts together and you will understand the I–V curve (red trace) of Figure 8.14 [12]. For zero bias, a short-circuit current (I_{sc}) will flow through the system. This current decreases when the forward voltage is increased, until a point is reached (V_{oc}) when no current flows through the system. The maximum power of the device is calculated by "sliding" I, V along the curve to maximize the area of the rectangle, as shown in Figure 8.14.

For a given illumination power (P), the conversion efficiency η is given by:

$$\eta = \frac{V_{max}I_{max}}{P} = \frac{V_{oc}I_{sc}FF}{P}, \qquad (8.2)$$

where FF, the filling factor, is introduced to simplify estimates: V_{oc} and I_{sc} are easily measured, and empirical relationships have been reported which reproduce FF reasonably well using V_{oc} as the only input parameter [14].

Further analysis of the process (see, for example, Ref. [13]) leads to quite interesting findings. V_{oc} and I_{sc} are related by:

$$V_{oc} = \frac{kT}{e} \ln \left(\frac{I_{sc}}{I_0} \right), \tag{8.3}$$

where e is the elementary charge and I_0 is the dark current. This current flows in the opposite direction as I_{sc} and must be minimized when engineering a device. From Eq. (8.3) it also follows that I_{sc} must be maximized in order to increase η. Alas, many factors contribute to I_{sc} so its maximization is not straightforward. Let us consider what happens to a photogenerated electron. The charge carrier will first move through the depletion region and then through the neutral semiconductor. Movement through the depletion region is governed by the drift length:

$$l_{drift} = \mu E \tau, \tag{8.4}$$

where E is the electric field built into the junction before application of the external voltage, μ is the mobility of the carrier and τ is the lifetime of the carrier. After crossing into the neutral region, the carrier will diffuse to the contact. The diffusion length is given by:

$$l_{diff} = \sqrt{D\tau}, \tag{8.5}$$

where D is the diffusion coefficient:

$$\frac{D}{\mu} = \frac{kT}{e}. \tag{8.6}$$

Eqs. (8.4) and (8.5) show that mobility and carrier lifetime are most relevant for the performance of solar cells. The parameters affecting lifetime and mobility are direct carrier recombination, recombination with defect-induced traps (both shallow and deep), and, for nanostructured systems, distance between the nanoparticles, presence and type of adsorbates (e.g., surfactants, or surface oxides).

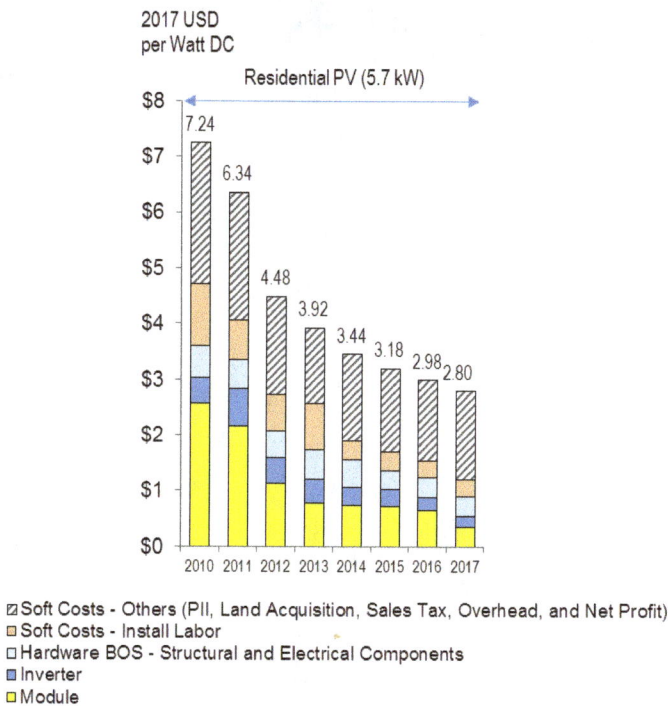

Figure 8.15: Cost breakdown for installation of residential, Si-based PV. Adapted with permission from [15]. © National Renewable Energy Laboratory.

8.5.1. *Why nanostructured photovoltaics?*

This is a good question; the answer to which has changed a bit over time. One of the reasons that is often brought about is weight. Nanostructured solar cells, and in particular, polymer-based solar cells, are lightweight; therefore they could save installation costs (Si-based PV are much heavier). However, installation costs of conventional, Si-based PV have decreased substantially over the years [15], as shown in Figure 8.15. Another reason that is often brought about is cost. Nanostructured solar cells (some at least) can literally be printed on plastic and are considerably cheaper than Si PV. However, one should note that the cost of Si modules has decreased by about 5 times in the last few years, as shown in Figure 8.15. In addition, nanostructured PVs are not as durable and as efficient as conventional PV modules, which also adds to cost. See Figure 8.16

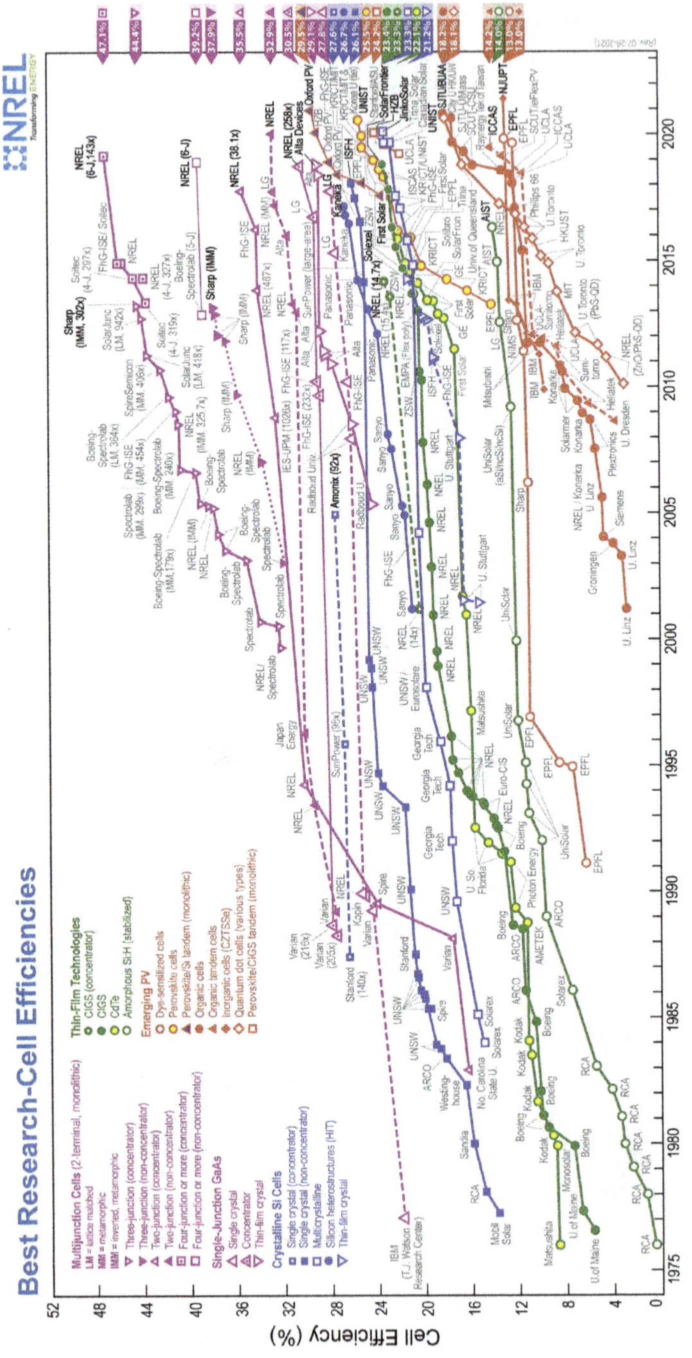

Figure 8.16: Efficiency of different types of solar cells. From [https://www.nrel.gov/pv/cell-efficiency.html]. © National Renewable Energy Laboratory.

for a chart of efficiencies: nanostructured solar cells have an efficiency <20% (typically 12-15%), while residential, Si-based PVs hover around 25%. One field, however, where nanostructured PV has an edge, is that of flexibility. Polymeric PV modules, for example, can be literally folded and put inside a bag. This feature is extremely relevant for military applications, among others.

8.5.2. *Dye-sensitized solar cells*

Following the discovery of water photodissociation on TiO_2 [16], research started to extend the spectral range of this all-important reaction. TiO_2 is a wide band-gap semiconductor ($E_{gap} \sim 3.2\,eV$); therefore UV light is necessary to use it in photocatalysis and PV. Other semiconductors with smaller band gap, such as CdS, would allow harvesting of visible light, but they are easily photocorroded. Early work [17] showed that the energy levels of dyes such as Rhodamine or porphyrines straddle the conduction band of TiO_2 and could therefore be used to inject electrons into TiO_2, as shown in Figure 8.17. The oxidized dye would be reduced to its initial, light-absorbing state by electrochemical reduction (typically using an iodine species), also as shown in Figure 8.17 [18]. While a powerful idea, a major technical hurdle of dye-sensitized solar cells (DSSC) emerges by analyzing the concept itself. That is, only dye molecules that are in direct contact with the TiO_2 electrode will contribute to the photocurrent. This means that the active layer is one molecule deep. Hence, the cell will nearly be transparent and the efficiency of the device will be low.

The issue of transparency and efficiency was addressed in Ref. [19], where a high surface area ZnO electrode was employed. The sensitizing dye was rose bengal, which absorbs in the green and also coordinates extremely well to the ZnO surface. The large surface area allowed for a light-absorbing region which was deeper than one monolayer, hence allowing a more efficient harvesting of light. Efficiencies were reported which were ~10 times higher than previous reports on low surface area substrates. The DSSC device was further refined by Grätzel, who in 1991 reported an efficiency of 11% [20]. The DSSC concept is very powerful, yet it suffers from durability issues. Organic dyes are rather rapidly degraded by the continuous

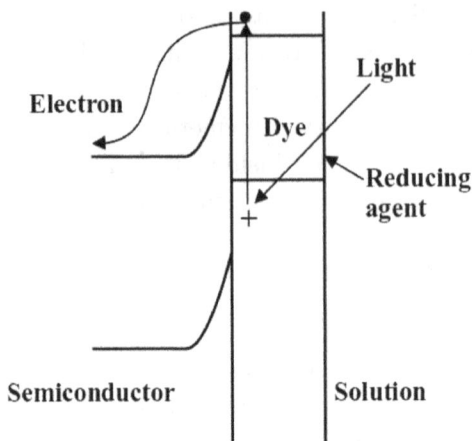

Figure 8.17: Energy levels of TiO₂ (left) and a sensitizing dye such as rhodamine. The circuit is closed electrochemically via a reducing agent.

redox cycles. In addition, the need for an electrolyte is an obstacle to portability and flexibility of DSSCs. For these reasons, current research focuses on solid state solar cells, be they DSSCs, polymeric, or hybrid organic-inorganic. In the following, we will focus on these two latter classes of nanostructued PV power generation schemes.

8.5.3. *Quantum dot-based PV*

One of the earliest implementations of quantum dots (specifically, PbS) as TiO₂ sensitizers was reported in 1995 [21]. PbS has a band gap in the infrared, and is therefore an excellent absorber of sunlight. Problem is, the energy levels of PbS do not straddle the conduction band of TiO₂, as shown in Figure 8.18. However, the levels of PbS quantum dots do fulfill the condition, as shown in Figure 8.18. The photoresponse of TiO₂ sensitized with PbS quantum dots is shown in Figure 8.19 and shows how these DSSCs could be used to harvest visible and near infrared light.

PbS-based solar cells and photodetectors have been further developed, among others, by the Sargent group in Toronto. Following the evolution of the research by this group is particularly instructive. In one of the earliest implementations [22], a film of PbS nanoparticles was deposited across Au electrodes, as

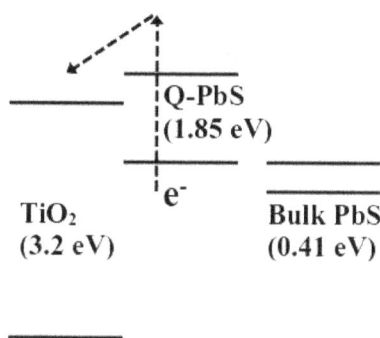

Figure 8.18:   Energy levels of TiO₂, bulk PbS and nano-PbS. Adapted with permission from [21]. © AIP Publishing.

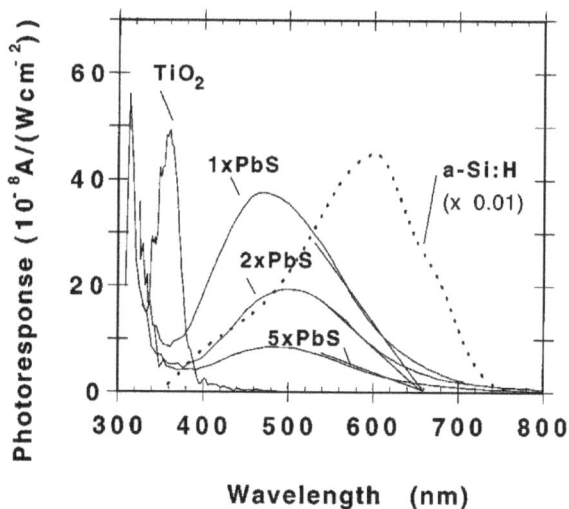

Figure 8.19:   Photoresponse of TiO₂ sensitized with PbS quantum dots. Adapted with permission from [21]. © AIP Publishing.

shown in Figure 8.20. The simplicity of the set-up allowed the authors to learn several things about the system. Three different types of quantum dots were used for the experiments. One set was capped with a with a short-chain capping agent, butylamine. A second set consisted of PbS particles which had been coagulated and then oxidized (neck-then oxidize). A third set of particles

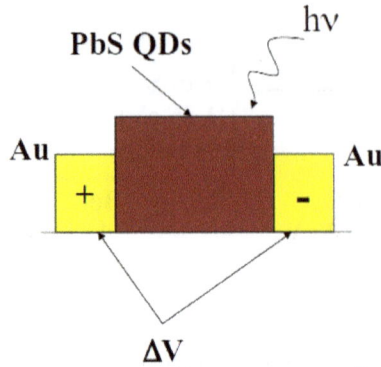

Figure 8.20: Schematic representation of a PbS based photodetector. A PbS film is deposited between two Au electrodes, with a potential difference applied across them.

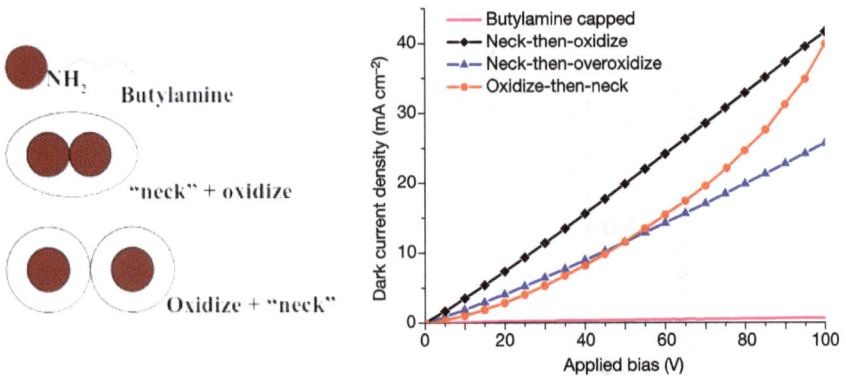

Figure 8.21: Left: structure of the particles used in the experiments. Right: measured photocurrent. Adapted with permission from [22]. © Springer Nature.

had been oxidized first, then coagulated (oxidized-then-neck). The types of nanoparticles and the corresponding photocurrents are reported in Figure 8.21. The photocurrents show that the most efficient systems were those which contained oxides. This result is counterintuitive, but was explained by measurements of the carrier lifetimes (τ). In oxidized systems, shallow defect traps are created. These traps capture the charge carriers. However, since they are shallow, the carriers can readily escape and reach the charge-collecting

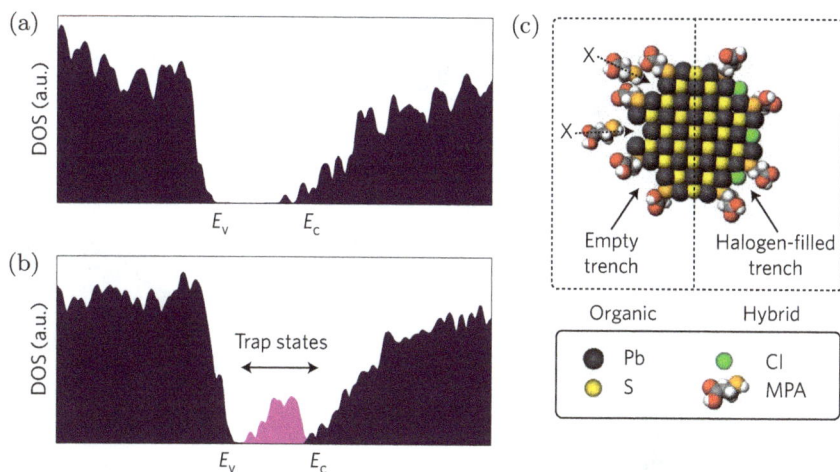

Figure 8.22:   (a) Density functional calculations of completely vs. (b) partially passivated PbS quantum dots. (c) schematic represenation of the passivation scheme involving thiols and halogens. Adapted with permission from [25]. © Nature Publishing Group.

electrodes. The increase in τ increases the device efficiency, as per Eq. (8.4). In follow-up work the group showed that oxidation state, type of ligand and ligand length affected τ, μ and the doping level. For example, τ varied from tens of ms in PbS capped with ethanethiol to seconds in systems where $PbSO_4$ was detectable [23]. This indicated that traps induced by ethanethiol were much shallower than those related to $PbSO_4$. The presence of deep traps was found to negatively affect exciton dissociation, which is fundamental in quantum dot PV [23]. It was also found that mobility increased when short ligands were employed. This result is not completely surprising. Charge carriers have to hop through nanoparticles, and the hopping probability increases for shorter interparticle distances [24].

The passivation strategy was further developed by adding halogen atoms post-synthesis [23]. The halogen atoms further reduced the number of traps, as shown by DFT calculations in Figure 8.22, and boosted the quantum efficiency of the devices to a record 7%.

The group later came up with an ingenious scheme for a tandem cell, capable of harvesting light from the near UV (350 nm) to the infrared ($\sim$1 µm) [26]. In this cell, quantum dots of different sizes

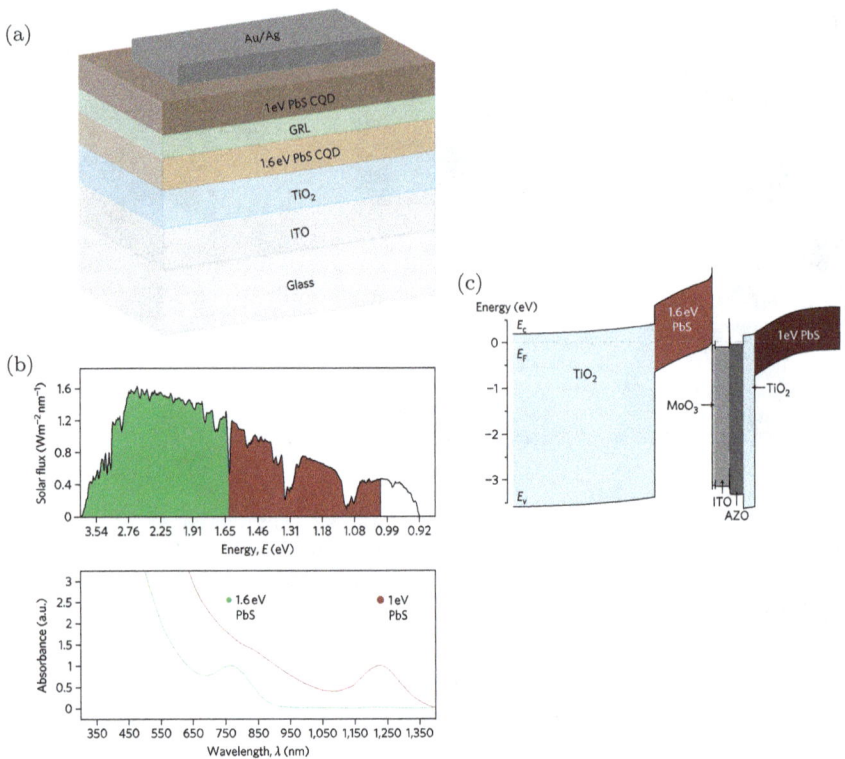

Figure 8.23: A tandem cell used to harvest light from UV to infrared. (a) Schematic cross-section. (b) spectral range covered and absorption spectrum of the PbS quantum dots employed to construct the device. (c) Energy levels of the system. Adapted with permission from [26]. © Nature Publishing Group.

are employed, the combination of whose absorption spectra covers a wide spectral range (Figure 8.23b). The upper part of the cell is a Schottky-type cell, where the quantum dots are in direct contact with a metal. The holes migrate towards the metal, while the electrons cross into TiO$_2$ and a series of oxides with decreasing energies of the conduction band. The second cell is a more "conventional" PbS-TiO$_2$ cell. The electrons generated by PbS in this cell are swept away by the TiO$_2$ electrode. The holes recombine with the electrons photogenerated in the Schottky cell and close the circuit.

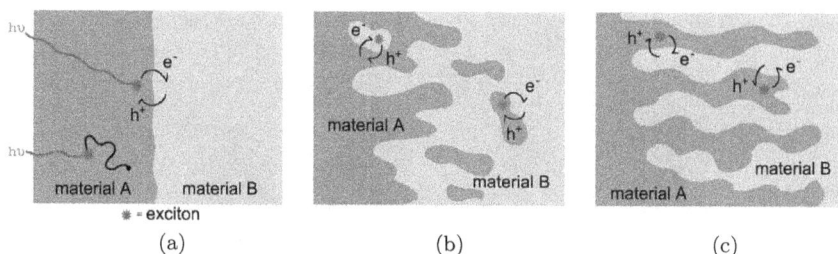

Figure 8.24: Possible geometries of conducting polymer interfaces. (a) planar interface. (b) Blend. (c) Interdigitated. Adapted with permission from [29]. © IOP Science.

8.5.4. *Conducting polymer-based PV*

Similar to the case of quantum dot-based PV, one has to keep in mind the diffusion length (<10 nm) [27] and the lifetime (∼300 ps) [28] of the exciton in conducting polymers. Excitons are most efficiently dissociated at interfaces, which means that polymers must be blended at the nanoscale in order to minimize exciton recombination [29]. Based on these considerations, interfaces like the one in Figure 8.24(a) are not expected to be very efficient. Only excitons generated near the interface will contribute to the photocurrent. This is of course disappointing, since such interfaces are easily fabricated, for example by sputtering, or by spin-coating. This limitation notwithstanding, solar conversion efficiencies of ∼1% were achieved in early work [30]. The interface of Figure 8.24(b), based on polymer blends, is more efficient and is also easy to fabricate. It was used by the group of R. Friend [31], who spin-coated a solution of two polyphenlene vinylene (PPV) derivatives, one hole- and one electron-transporting. The two polymers phase separated on the substrate, as shown in Figure 8.25. The corresponding solar cells had a quantum efficiency of ∼6%. A third approach to fabricate nanostructured interfaces is represented by interdigitated polymer blends, shown in Figure 8.24(c). Such interfaces are most efficient; yet they require electrochemical methods for their preparation, which limits throughput and device area.

An interesting variation of the polymer blend approach was developed by the group of Scherf [32]. In this approach solar cells are

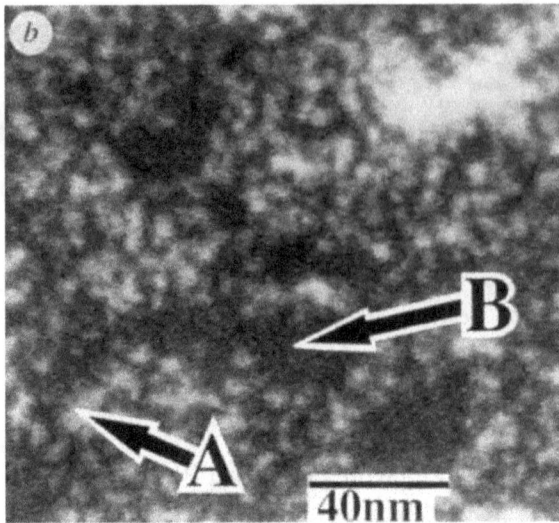

Figure 8.25: Bright field transmission electron micrograph showing polymer blending in a polymer-based solar cell. Staining was used to enhance the contrast between the two polymer types. Adapted with permission from [31]. © Nature Publishing Group.

fabricated using nanoparticles of blended polymers. These solar cells are up to a factor 4 more efficient than solar cells fabricated by blending nanoparticles of the individual constituents, as shown in Figure 8.26. Following a similar line of thought, the same group demonstrated that solar cells that included spherical quantum dots and semiconductor nanowires overperformed systems of the individual components, as shown in Figure 8.27 [33]. The performance improvement was attributed to the formation of percolation pathways by the nanorods.

The body of evidence reported in Sections 8.5.3 and 8.5.4 shows therefore that an ideal PV cell should have long recombination times, high mobilities, and a limited number of surface defects. Figure 8.28 summarizes the different processes (and losses) that happen in an organic PV cell.

Among these processes, free carrier recombination (also called non-geminate recombination) is responsible for most photocurrent

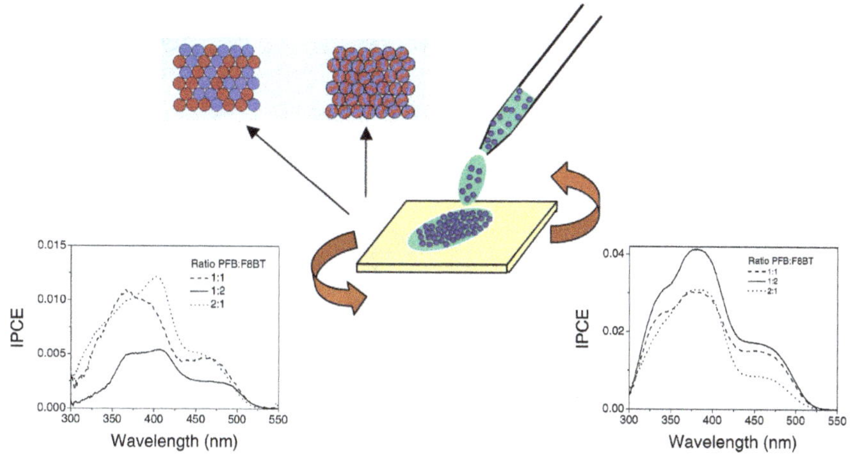

Figure 8.26: Quantum efficiency of solar cells based on blends of polymer nanoparticles (left) and of nanoparticles of blended polymers (right). Adapted with permission from [33]. © American Chemical Society.

losses. These losses (I_{NG}) can be expressed as [34]:

$$I_{NG}(V) = e\frac{d \cdot n}{\tau(n)}, \qquad (8.7)$$

where n is the carrier density, $\tau(n)$ is the recombination time, and d the cell thickness. τ depends on the carrier density but also on the system, as shown in Figure 8.29 [35].

Several empirical relations have been derived for n and τ. Popular choices are:

$$n = n_0 e^{\gamma V_{oc}}, \\ \tau - \tau_0 n^{-\lambda}, \qquad (8.8)$$

where n_0, τ_0, λ and γ are fitting parameters. In systems with low carrier mobility, the loss of carrier density through non-geminate recombination can be modelled by a Langevin relation for bimolecular recombination:

$$\frac{dn}{dt} = \frac{-n}{\tau(n)} = -\gamma_L n^2, \qquad (8.9)$$

(a)

(A)   (B)   (C)

(b)

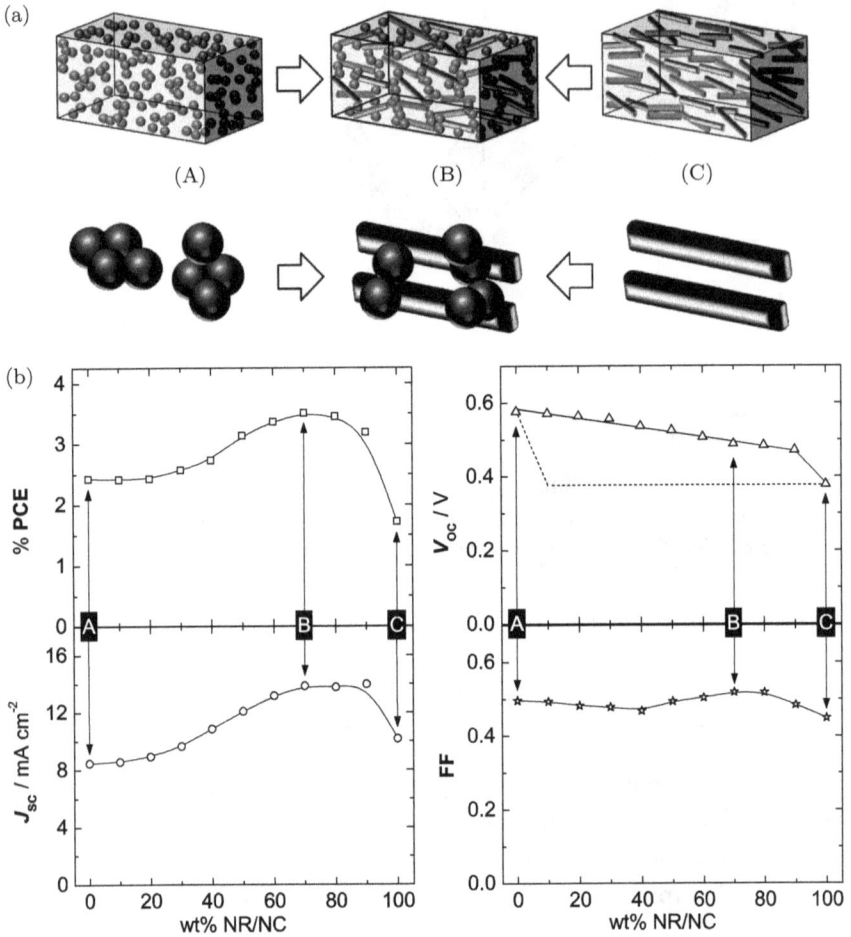

Figure 8.27:   (a) Schematic illustrations of light-absorbing nanoparticles embedded in a polymer matrix. A quantum dots, B quantum dots and nanorods, C
nanorods. (b) Conversion efficiency (squares), short-circuit current (circles), Voc
(triangles), and FF (asterisks) of devices where the QD/NR weight ratio was
varied while the polymer content was kept constant at 10 wt%. Adapted with
permission from [32]. © Nature Publishing Group.

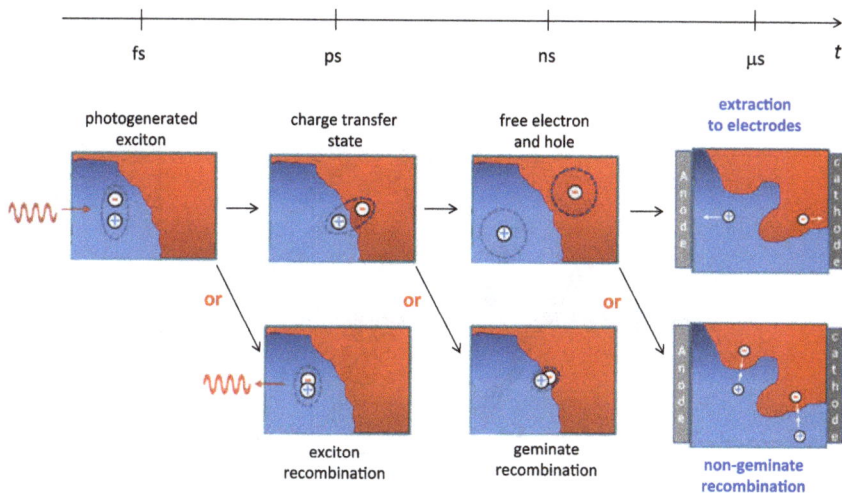

Figure 8.28: Processes leading to free carrier generation and extraction. Starting from left: absorption of light generates a short-lived exciton. The exciton can recombine or diffuse to-, and dissociate at- the donor-acceptor heterojunction. Due to the low permittivity of polymers, a strong electric field is present even after dissociation. The charges can overcome this field to form free carriers, or recombine geminately. Charges that survive this stage can be extracted at the electrodes, even though non-geminate recombination remains an issue. The order of magnitude of the time scales for these processes is reported at the top. Adapted with permission from [34]. © Springer.

where

$$\gamma_L = \frac{e(\mu_e + \mu_h)}{\epsilon_0 \epsilon_r}. \tag{8.10}$$

When engineering a device, one should consider that measurement of recombination times and mobilities is tricky. Polymers are disordered systems; hence results depend strongly on synthesis and processing conditions. In addition, the most frequently employed experimental set-ups are based on pump-probe laser techniques. These techniques require high excitation densities, which may not reflect the "normal" operating conditions of solar cells. More gentle methods have been recently developed, which are based on modulation of laser pulses and collection bias [35]. Use of these methods has shown that the parameter γ in Eq. (8.8) can change by up to 10 times by simply annealing a device. Annealing also improves

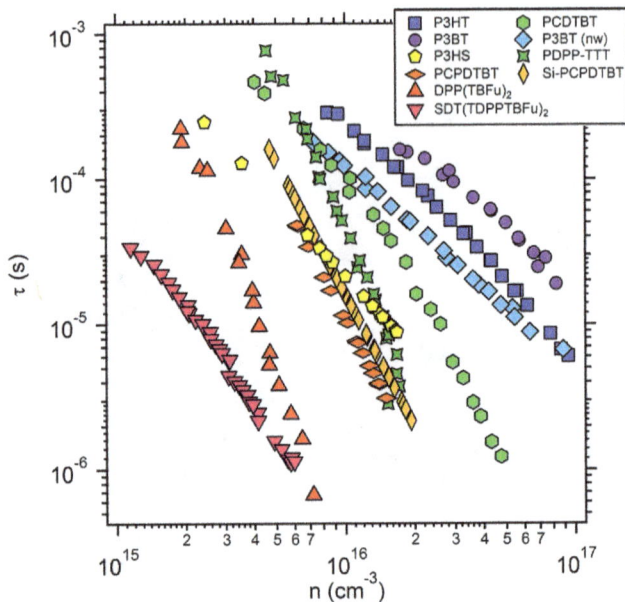

Figure 8.29: Dependence of carrier lifetime on carrier density for a series of systems. Adapted with permission from [35]. © American Chemical Society.

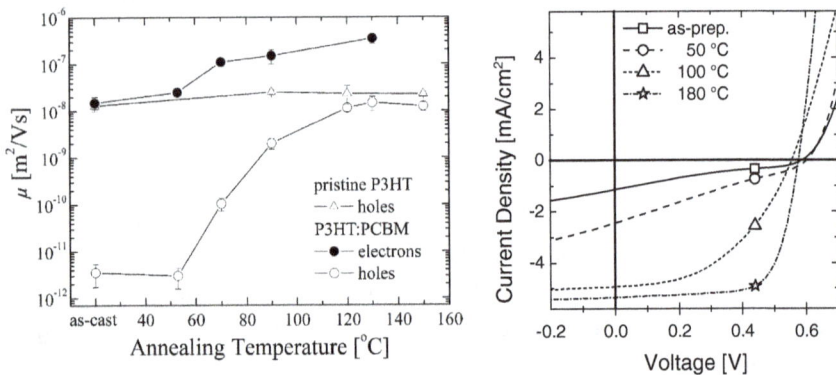

Figure 8.30: Effect of annealing temperature on organic PV performance. Left: increase in carrier mobility. Right: increase in efficiency. Figures adapted with permission from [36, 37]. © Wiley-VCH.

mobilities [36], and with that, device performance [37], as shown in Figure 8.30.

References

1. A. G. MacDiarmid, R. J. Mammone, R. B. Kaner, and S. J. Porter, *Phil Trans. R. Soc. Lond. A* **314**, 3–15 (1985).
2. T. Le, Y. Kim, and H. Yoon, *Polymers* **9**, 150 (2017).
3. P. Chandrasekhar, "Conducting Polymers, Fundamentals and Applications", 2nd Ed., Springer, Berlin 2018.
4. D. C. Bott, "Structural basis for semiconducting and metallic polymers". In: Skotheim, T.A. (ed.) "Handbook of conducting polymers", vol. 2, p. 1191. Marcel Dekker, Inc., New York (1986).
5. N. F. Mott, *Philosophical Magazine. Informa UK Limited.* **19**(160): 835–852 (1969).
6. A. J. Epstein, "AC conductivity of polyacetylene: Distinguishing mechanisms of charge transport". In: Skotheim, T.A. (ed.) "Handbook of conducting polymers", vol. 2, p. 1047. Marcel Dekker, Inc., New York (1986).
7. L. K. Werake, J. G. Story, M. F. Bertino, S. K. Pillalamarri, and F. D. Blum, *Nanotechnology* **16**, 2833–2837 (2005).
8. S. Pillalamarri, F. D. Blum, A. T. Tokuhiro, and M. F. Bertino, *Chem. Mater.* **17**, 5941–44 (2005).
9. Z. F. Li, F. D. Blum, M. F. Bertino, C. Kim, *Sensors and Actuators B* **161**, 390–395 (2012).
10. N. C. Greenham, F. Cacialli, D. D. C. Bradley, R. H. Friend, S. C. Moratti, and A. B. Holmes: "Cyano-derivative of poly (P-Phenylene Vinylene) for use in thin-film light-emitting diodes". In: Electrical, optical, and magnetic properties of organic solid state materials, vol. 328, p. 351. Materials Research Society, Pittsburgh (1994)
11. H. Mattoussi, L. H. Radzilowski, B. O. Dabbousi, E. L. Thomas, M. G. Bawendi, and M. F. Rubner, *J. Appl. Phys.* **83**, 7965 (1998).
12. J. Tang and E. H. Sargent, *Adv. Mater.* **23**, 12–29 (2011).
13. S. M. Sze, "Semiconductor Devices, Physics and Technology", Wiley, New York, 1985.
14. M. A. Green, *Solid State Electronics*, **24**, 788 (1981).
15. R. Fu, D. Chung, T. Lowder, D. Feldman, K. Ardani, and R. Margolis, "US solar photovoltaic system cost benchmark: Q1 2016", National Renewable Energy Laboratory.
16. A. Fujishima and K. Honda, *Nature* **238**, 37 (1972).
17. H. Gerischer, *Photochem. Photobiol.* **16**, 243 (1972).
18. L. R. Faulkner, "Electrochemical Methods: Fundamentals and Applications" 1st edition, Wiley, New York, 1980.
19. H. Tsubomura, M. Matsumura, Y. Nomura, T. Amamiya, *Nature* **261**, 402 (1976).

20. B. O'Reagan and M. Grätzel, *Nature* **353** (1991) 737.

21. P. Hoyer and R. Könenkamp, *Appl. Phys. Lett.* **66**, 349–351 (1995).

22. G. Konstantatos, I. Howard, A. Fischer, S. Hoogland, J. Clifford, E. Klem, L. Levina and E. H. Sargent, *Nature* **442**, 180 (2006).

23. G. Konstantatos, L. Levina, A. Fischer, E. H. Sargent, *Nano Letters*, **8**, 1446 (2008).

24. D. Aaron R. Barkhouse, Andras G. Pattantyus-Abraham, Larissa Levina, and Edward H. Sargent, *ACS Nano*, **2**, 833 (2008).

25. A. H. Ip, S. M. Thon, S. Hoogland, O. Voznyy, D. Zhitomirsky, R. Debnath, L. Levina, L. R. Rollny, G. H. Carey, A. Fischer, K. W. Kemp, I. J. Kramer, Z. Ning, A. J. Labelle, K. W. Chou, A. Amassian, and E. H. Sargent, *Nature Nanotechnology* **7**, 577 (2012).

26. X. Wang, G. I. Koleilat, J. Tang, H. Liu, I. J. Kramer, R. Debnath, L. Brzozowski, D. A. R. Barkhouse, L. Levina, S. Hoogland, and E. H. Sargent, *Nature Photonics Letters* **5**, 480 (2011).

27. O. Mikhnenko and P. Blom, *Energy and Environmental Science*, **8**, 1867–1888 (2015).

28. D. Moses, *Synthetic Metals* **125**, 93–98 (2002).

29. A. B. F. Martinson, A. M. Massari, S. J. Lee, R. W. Gurney, K. E. Splan, J. T. Hupp, and S. T. Nguyen, *J. Electrochem. Soc.* **153**, A527 (2006).

30. C. W. Tang, *Appl. Phys. Lett.* **48**, 183 (1986).

31. J. J. M. Halls, C. A. Walsh, N. C. Greenham, E. A. Marseglia, R. H. Friend, S. C. Moratt, and A. B. Holmes, *Nature* **376**, 498 (1995).

32. T. Kietzke, D. Neher, K Landfester, R. Montenegro, R. Güntner, and U. Scherf, *Nature Materials* **2**, 408 (2003).

33. T. Kietzke, D. Neher, M. Kumke, R. Montenegro, K. Landfester, and U. Scherf, *Macromolecules* **37**, 4882–4890 (2004).

34. S. Ludwigs, Ed., "P3HT Revisited — from Molecular Scale to Solar Cell Devices", Springer, Heidelberg, 2014.

35. D. Credgington and J. R. Durrant, *J. Phys. Chem. Lett.* **3**, 1465–1478 (2012).

36. V. D. Mihailetchi, H. X. Xie, B. de Boer, L. J. A. Koster, and P. W. M. Blom, *Adv. Funct. Mater.* **16**, 699–708 (2006).

37. S. T. Turner, P. Pingel, R. Steyrleuthner, E. J. W. Crossland, S. Ludwigs, and D. Neher *Adv. Funct. Mater.* **21**, 4640–4652 (2011).

Index

www.ingramcontent.com/pod-product-compliance
Lightning Source LLC
Chambersburg PA
CBHW060255220326
41598CB00027B/4108